About Island Press

Since 1984, the nonprofit organization Island Press has been stimulating, shaping, and communicating ideas that are essential for solving environmental problems worldwide. With more than 1,000 titles in print and some 30 new releases each year, we are the nation's leading publisher on environmental issues. We identify innovative thinkers and emerging trends in the environmental field. We work with world-renowned experts and authors to develop cross-disciplinary solutions to environmental challenges.

Island Press designs and executes educational campaigns in conjunction with our authors to communicate their critical messages in print, in person, and online using the latest technologies, innovative programs, and the media. Our goal is to reach targeted audiences—scientists, policymakers, environmental advocates, urban planners, the media, and concerned citizens—with information that can be used to create the framework for long-term ecological health and human well-being.

Island Press gratefully acknowledges major support of our work by The Agua Fund, The Andrew W. Mellon Foundation, The Bobolink Foundation, The Curtis and Edith Munson Foundation, Forrest C. and Frances H. Lattner Foundation, The JPB Foundation, The Kresge Foundation, The Oram Foundation, Inc., The Overbrook Foundation, The S.D. Bechtel, Jr. Foundation, The Summit Charitable Foundation, Inc., and many other generous supporters.

The opinions expressed in this book are those of the author(s) and do not necessarily reflect the views of our supporters.

Replenish

Replenish

The Virtuous Cycle of Water and Prosperity

Sandra Postel

ISLAND PRESS is a trademark of The Center for Resource Economics.

Library of Congress Control Number: 2017936498

All Island Press books are printed on environmentally responsible materials.

Manufactured in the United States of America
10 9 8 7 6 5 4 3 2 1

Keywords: drought, floods, wildfire, climate change, wastewater, irrigation, dams, rain capture, regenerative agriculture, restoration.

To Virginia

Contents

CHAPTER 1

Water Everywhere and Nowhere

In water that departs forever and forever returns,
we experience eternity.
—*Mary Oliver*

As I wound my way up Poudre Canyon in northern Colorado, the river flowed toward the plains below, glistening in the midday sun. It ran easy and low, as it normally does as the autumn approaches, with the snowmelt long gone. I was struck by the canyon's beauty, but also by the blackened soils and charred tree trunks that marred the steep mountains all around. They were legacies, I realized, of the High Park Fire that had burned more than 135 square miles (350 square kilometers) of forest during the previous year's drought. It was September 7, 2013, and my family and I were heading to my niece's wedding. Tara and Eric had chosen a spectacular place for their nuptials—Sky Ranch, a high-mountain camp not far from the eastern fringe of Rocky Mountain National Park. As we escorted my elderly parents down the rocky path to their seats, I noticed threatening clouds moving in. They darkened as the preacher delivered his homily. Please cut it short and marry them, I thought to myself, before we all get drenched.

The rains held off just long enough. But that day's brief shower was a prelude to a deluge of biblical proportions that began four days later. A storm system stalled over the Front Range and in less than a week dumped nearly a year's worth of precipitation in some areas. The Poudre—short for Cache la Poudre—flooded bigger than it had since 1930. The torrential rains washed dead tree trunks down the hillsides into the raging river below. One canyon resident wrote that the blackened logs "looked like Tinker Toys amid the river's mad rush."[1]

The threefold punch of drought, fire, and flood wreaked even worse havoc in neighboring mountain canyons, including that of the Big Thompson, a river renowned for the devastating flood of 1976. While that flood took 144 lives, it was relatively localized. This 2013 flood was vast, covering most of Colorado's Front Range and affecting not only high-elevation towns from Boulder to Estes Park—a number of which experienced a 1-in-500-year storm—but the heavily populated plains from Colorado Springs north to Fort Collins. Though by no means the deadliest, with eight lives lost, it became one of the costliest flood events in Colorado's history. It triggered 1,300 landslides, damaged some 19,000 homes and commercial buildings, required the evacuation of more than 18,000 people, damaged 27 state dams (and completely took out a handful of "low-hazard" dams), and damaged or destroyed 50 bridges and 485 miles (780 kilometers) of roads. Losses were estimated to total some $3 billion.[2]

Floods of this magnitude, while rare overall, are completely unexpected in Colorado in the very late summer. In river systems fed by melting snows, the biggest floods normally occur in the spring, as temperatures warm and snowmelt pours into headwater streams and the rivers they feed. Intense summer thunderstorms occasionally create localized flooding in July or August, but by September rivers are typically running low, just as the Poudre was when I drove up the canyon.

Brad Udall, a water and climate expert at the University of Colorado in Boulder, whose house sits just 30 feet (9 meters) from a creek that's normally dry in September, saw the creek turn into a raging stream. "This was a totally new type of event," Udall told *National Geographic*, "an early-fall, widespread event during one of the driest months of the year."[3]

So often these days water seems to be nowhere and everywhere all at once. The wild weather of 2015 became almost legendary, even before the year was over. With raging floods in Latin America, the US Midwest, and the United Kingdom, and withering droughts in eastern and southern Africa, most of California and southeastern Brazil, terms such as *anomalous*, *historic*, and *epic* dominated the weather lexicon. US scientists determined that during one rare October rainstorm 17 streams in the US state of South Carolina broke records for peak flow. According to the United Nations, two years of drought left nearly 1 million African children suffering from acute malnutrition, and millions more at risk from hunger, water shortages, and disease.[4]

Although the weather phenomenon known as El Niño became the go-to explanation for the global turmoil that year, this periodic event was not fully to blame. The El Niño came atop long-term warming trends that are fundamentally altering the movement of water across the planet. The earth was hotter in 2016 than since record keeping began in 1880. The previous record was 2015, which itself had beaten the previous record of 2014 by a considerable margin. For the contiguous United States, 2016 marked the twentieth consecutive year that the annual average temperature was higher than the twentieth-century average.[5]

As air warms, it expands, which allows it to hold more moisture. This, in turn, increases evaporation and precipitation, which generally makes dry areas drier and wet areas wetter. If disasters related to droughts, floods, and other extreme weather seem more common globally, it's because they are: according to a United Nations study, between 2005 and 2014, an average of 335 weather-related disasters occurred per year, nearly twice the level recorded from 1985 to 1995.[6]

If we don't adapt to these new circumstances, a future of more turmoil is bound to unfold. The 6,457 floods, storms, droughts, heat waves, and other weather-related events that occurred over the last two decades caused 90 percent of disasters during that period. Those disasters claimed more than 600,000 lives and cost more than $1.9 trillion, according to the UN study. The countries hit with the highest number of disasters over the twenty-year period were the United States, with 472, and China, with 441, followed by India, the Philippines, and Indonesia.

Meanwhile, extreme weather is also affecting our food supply. A team of Canadian and UK scientists found that from 1964 to 2007 droughts and heat waves had each slashed the production of cereals by about 10 percent—and by 20 percent in the more-developed countries. Altogether, the loss was estimated at 3 billion tons.[7]

Leaders in business and government are beginning to take notice. More than 90 percent of companies in the S&P Global 100 Index see extreme weather and climate change impacts as current or future risks to their business.[8] At its annual gathering in Davos, Switzerland, in 2016, the World Economic Forum—which counts among its members heads of state, chief executive officers, and civic leaders—declared water crises to be the top global risk to society over the next decade. Next on the list were the failure to mitigate and adapt to climate change, extreme weather events, food crises, and profound social instability.[9] All five threats are intimately connected to water. Guarding against each requires a new understanding of our relationship to freshwater—and a new way of thinking about how we use, manage, and value it.

～

Water is unlike any other substance. It is always on the move—falling, flowing, swirling, infiltrating, melting, condensing, evaporating—and all the while knitting the vast web of life together. Through its endless circulation, water connects us across space and time to all that has come before and all that is yet to be. Our morning coffee might contain molecules the dinosaurs drank.

This profound connection is created by one of the most mysterious and underappreciated of Earth's natural phenomena: the water cycle. Those fifth-grade textbook diagrams never quite do it justice. We see the labels of water stocks and flows and the arrows signaling movement from sea to air to land, but never really grasp the magic wrought by two atoms of hydrogen uniquely bonded to one of oxygen. Water is the only substance that can naturally exist as a liquid, gas, or solid at normal Earth temperatures.

With hydrogen from the primordial Big Bang and oxygen from early stardust, water was born. Infant Earth, hot as Hades, was enveloped in water vapor, but it took a billion or more years of cooling before that vapor could condense and fall to the young planet's surface as rain. Liquid water has wetted Earth for at least three billion years. Today, that stock of water is finite, except perhaps for minute additions from so-called cosmic snowballs—small comets made of water that smash into the earth.

This finite supply circulates over vastly different scales of time and space. Some water molecules get trapped ultradeep within the earth, remain there for millennia, and then suddenly burst into the atmosphere through an erupting volcano. Others reside close to the earth's surface, changing back and forth between liquid and vapor as they evaporate from a lake, condense into a cloud, and fall as rain to join a river as it flows to the sea. From there, they evaporate again, and the cycle continues. Still other molecules remain trapped for centuries in glacial ice until they melt to replenish a mountain meadow and the groundwater below. "Whenever you eat an apple or drink a glass of wine," writes astrophysicist and author Robert Kandel, "you are absorbing water that has cycled through the atmosphere thousands of times since you were born. But you are also absorbing some water molecules that have only been out in the open air for a few days or weeks, after tens or hundreds of millions of years beneath the Earth's crust."[10]

Almost all the water on Earth—97.5 percent—resides in the ocean and is too salty to drink or to irrigate most crops. Of the remainder, about two-thirds is locked up in glaciers and ice caps. Only a tiny share of Earth's water—less than one one-hundredth of one percent—is both fresh and continuously renewed by the solar-powered global water cycle.

Each year, the sun's energy lifts nearly 500,000 cubic kilometers (132 quadrillion gallons) of water from the earth's surface—86 percent from the oceans and 14 percent from the land.[11] An equal amount falls back to Earth as rain, sleet, or snow, but, fortunately for us, not in the same proportions. Wind and weather transfer about 9 percent of the vapor lifted from the sea over to the land. This net addition of about 40,000 cubic kilometers combines with the 70,000 lifted from the land and its vegetation each year to create our total annual renewable water supply: 110,000 cubic kilometers (29 quadrillion gallons). The 40,000 cubic kilometers distilled and transferred from the oceans to the land makes its way back to the sea through rivers and shallow groundwater—what hydrologists call "runoff"—completing the global cycle and balancing nature's water accounts.[12]

That runoff is what we tap to irrigate crops, supply water to our homes and businesses, manufacture all of our material goods, and run turbines to generate electricity. It is also the water supply for all the fish, birds, insects, and wildlife that depend on rivers, streams, and wetlands for their habitats. Although the water cycle delivers that runoff each year, water is not always where we need it when we need it. Nature's water deliveries are often poorly matched with where people live or farmers find it best to grow crops. Today, for example, China is home to 19 percent of the world's population, but only 7 percent of global runoff.[13]

Although we speak of a global cycle, water circulates at many scales. Consider, for example, the tomato plant in your garden. Through its roots, it takes up moisture from the soil supplied by rain (and perhaps your extra watering), keeps some of it to fill its growing stems and leaves, and releases the rest in the form of vapor back to the atmosphere through openings in its leaves. Once aloft it may condense and fall again as rain. Similarly with the human body, 60 percent of which is water. We take water in through food and drink, rehydrate, and then release water back to the environment either in liquid form through our urine or in vapor form through our breath and the evaporation of our sweat. All terrestrial plant and animal life participates in the cycling of water.

During the ten thousand years since *Homo sapiens* opted for settled agriculture over its earlier hunter-gatherer existence, human activities have increasingly altered local, regional, and, more recently, global water cycles. Among the earliest people to do so on a substantial scale were the Sumerians, who migrated out of the Mesopotamian highlands some 5,500 years ago and settled in the lowland plains of the Fertile Crescent, in what is now southern Iraq. Their new locale was sunnier and, in that way, better for growing crops, but it lacked rainfall at critical times during the growing season. So the Sumerians constructed canals to transport water from the Euphrates River to their fields, and as a result became the first society in the world based on irrigation.[14]

Little did the Sumerians know, however, that this alteration of water's natural journey would be their undoing. The reason was not the water war 4,500 years ago between the two Mesopotamian city-states of Lagash and Umma. It was salt. The river water helped their wheat to grow, but once it transpired through the plants and evaporated from their fields, it left its natural salts behind—salts the Euphrates River would have otherwise carried to the Persian Gulf. As the salts accumulated in the soil, their wheat yields declined. The Sumerian farmers tried growing barley, a more salt-tolerant crop, but eventually those yields declined as well. When the land could no longer produce enough food, the people of Sumer packed up and headed north, leaving a salty wasteland behind.[15]

Since those early experiments of hydraulic manipulation, the scale and variety of human interventions in water's natural flow through the landscape have grown tremendously. By the second century BC, the Han dynasty in China was building earthen dams 30 meters (98 feet) high. But it was really in the mid-nineteenth century with advances in hydraulics, fluid mechanics, civil engineering, and other applied sciences that the construction of large-scale water infrastructure took off. In 1885, the British began remaking the Indus River Valley in colonial India into a massive irrigation network for the production of wheat. Although plagued by the scourge of soil salinity, just as the Sumerian lands had been long before, the Indus scheme eventually became the world's largest contiguous irrigation network, spanning 14 million hectares (35 million acres), an area a bit larger than the country of Costa Rica.[16]

Late nineteenth- and early twentieth-century scientific advances coincided with an evolving utilitarian philosophy that nature could be fundamentally transformed. Samuel P. Hays, in his 1959 book *Conservation and the Gospel of Efficiency*, described how, just after 1900, large-scale river development "suddenly captured the imagination" of conservation leaders. They grasped that "flood waters, now wasted, could, if harnessed, aid navigation, produce electric energy, and provide water for irrigation and industrial use."[17] In 1908 Winston Churchill stood on the shore of Africa's Lake Victoria, watching its waters spill over Owen Falls into the White Nile, and later reflected on the experience: "So much power running to waste . . . cannot but vex and stimulate the imagination. And what fun to make the immemorial Nile begin its journey by diving into a turbine."[18]

In that same vein, geologist and inventor William J. McGee, who held prominent US government and scientific positions during the late nineteenth and early twentieth centuries, wrote with prescience in 1909 that "the conquest of nature is now extending to the waters on, above, and beneath the surface. The conquest will not be complete until these waters are brought under complete control."[19]

These aspirations came to fruition in 1935 with the completion of the architecturally stunning Hoover Dam (originally named Boulder Dam) on the Colorado River in the southwestern United States. Hoover gave rise to the age of super dams and a whole new degree of control over water. US engineers actively exported their dam-building knowledge and expertise to other countries, and within decades arid lands around the world were open for business. With access to water, cities and farms spread like mushrooms in damp woods. Large reservoirs and tall levees offered a degree of flood control that encouraged farms and cities to locate in river floodplains, where they had access to rich soils and shipping corridors. Turbines affixed to big dams churned out electricity that propelled economies forward. In a speech in July 1954, India's prime minister, Jawaharlal Nehru, referred to dams as "the temples of modern India."[20]

The construction of these "modern temples" proceeded at a rapid clip. During the last half of the twentieth century, the nations of the world built an average of two large dams *a day*. As the twenty-first century dawned, some 45,000 large dams—those 15 meters (49 feet) or higher—blocked the world's rivers. China was also proceeding with the world's biggest river diversion scheme to transfer water more than 1,000 kilometers (600 miles) from the Yangtze River in the south to the drier north. Farmers around the globe were pumping vast quantities of groundwater to the surface to irrigate their fields and boost their harvests. By then, hydropower accounted for 19 percent of global electricity use. Populations were growing fastest in some of the world's driest places.

It is hard to say whether the growing demand for water and development during the last half of the twentieth century was the cause or consequence of this massive hydraulic engineering. In some ways, big water infrastructure has the same effect that commercial advertising does—it creates demand for its product: if you build it, they will come—and consume. To no small degree, that is what happened.

Around the world, humanity's thirst for water grew along with the big dams, canals, and material consumption made possible by control over water. It takes water to make everything—from computers to burgers and blue jeans. Because crops transpire so much water as they grow in farmers' fields, our diets are particularly water intensive. In fact, every day we "eat" a thousand times more water than we drink. A delicious margherita pizza takes about 1,250 liters (330 gallons) of water to make, most of it consumed during the growth of the tomatoes and the feed for the dairy cows that are milked to make the mozzarella cheese. Likewise, a cup of coffee requires some 130 liters (34 gallons), the majority of it transpired by the coffee bean plants. Our clothing consumes a great deal of water, as well—including some 2,500 liters (660 gallons) to make a simple cotton shirt. On any given day we are likely "wearing" more than 15,000 liters' (roughly 4,000 gallons') worth of water.[21]

Totaling it all up, it takes about 7,500 liters (nearly 2,000 gallons) of water a day to keep the average American lifestyle afloat. About half of that water is hidden in our diets, a third in the energy we use for travel and to heat and light our homes, 5 or 10 percent in the material goods we buy, and the remaining 5 or 10 percent for household activities, such as bathing, cooking, and watering our gardens and lawns.[22]

In part because Americans are quite carnivorous, and meat often (although not always, as we'll see later) takes a lot of water to produce (think about irrigating the grain to feed the cows), the typical American's water footprint is twice the global average. But humanity's collective global footprint is large, as well, and growing, as world population expands by some 244,000 people per day and many millions move up the income ladder every year.[23]

Researchers Arjen Hoekstra and Mesfin Mekonnen in the Netherlands have made the most detailed estimate to date of the scale and patterns of humanity's water consumption. Using a high level of spatial resolution, they tabulated all the water from both rainfall and irrigation that's consumed in making goods and services for the global population. They also added in the volume of water needed to assimilate the pollution generated along the way. When they calculated the annual average global footprint for 1996–2005, the most recent ten-year period for which the necessary data were available, their result was a whopping 9,087 billion cubic meters (2,400 trillion gallons) per year. That's more than 500 Colorado Rivers.[24]

In some ways it's hard to imagine our world of 7.5 billion people and $80 trillion in annual goods and services without water engineering—dams to store water, canals to move it around, and vast pumps to tap underground supplies. But it's equally hard to imagine continuing down this same path. Dams and reservoirs now intercept about 35 percent of river flows as they head toward the sea, up from 5 percent in 1950. Reservoirs have trapped more than 100 billion tons of sediment that rivers would otherwise have carried to the sea to replenish the coasts. As a result, productive deltas from the Mississippi to the Nile are losing ground to the sea, and barrier islands no longer offer coastal properties the same degree of protection from hurricanes and storms.[25]

Large dams have directly displaced some 40–80 million people and threatened the livelihoods of nearly 500 million more who depend on fishing, grazing, and farming activities contingent on the natural flows of rivers.[26] In addition, the diversity of life in freshwaters is undergoing a massive contraction that will surely worsen. The projected extinction rate for freshwater animal species in North America is about five times greater than that projected for the region's terrestrial species.[27] A 2016 article in the journal *Science* by forty researchers from eight countries warns that the more than 450 dams planned or under construction in the Amazon, Congo, and Mekong River basins threaten up to one-third of the world's freshwater fish species, many of which are found nowhere else. In the Amazon basin alone, where 334 additional dams are planned or proposed, some 64 percent of the basin's 2,320 fish species are endemic to that region.[28]

The blocking and diverting of rivers is not the only way we have broken nature's water cycle in the pursuit of economic progress. Groundwater depletion has more than doubled since 1960 and is now widespread in many of the world's most important food-producing regions. Watersheds shorn of trees no longer capture, store, and purify rainwater. Rivers bounded by levees rush floodwaters rapidly down their channels, increasing downstream flood risks. The disconnection of rivers from their floodplains has reduced groundwater recharge, the natural cleansing of river water, as well as habitats crucial for birds and fish. Rivers bearing high loads of nitrogen from fertilizer runoff that wetlands might otherwise absorb instead contribute to the creation of more than 400 low-oxygen dead zones in coastal bays and estuaries around the world. Soils depleted of microbes and organic matter due to poor land-use practices no longer hold moisture for plants and crops to draw upon during dry spells. And the impermeable pavement that coats urban and suburban landscapes causes storm water to run rapidly off the land, resulting in flooded streets and homes and polluted creeks and bays.[29]

For most of the last two centuries, these downsides of large-scale water engineering seemed to pale in comparison with the benefits. As long as progress was measured by the growth in populations served, hectares irrigated, and kilowatt-hours generated, the construction of big dams, canals, turbines, and pumps was deemed to serve humanity well. But the scales are tipping in the other direction as concerns about the costs, risks, fairness, and sustainability of this hydrologic engineering mount.

First, many regions have already overshot the sustainable limits of their water supply. An unsettling number of large rivers—including the Colorado and Rio Grande in the US Southwest, the Ganges and Indus in South Asia, the Amu Darya in central Asia, the Yellow in northern China, the Nile in northeastern Africa, and the Murray in southeastern Australia—are now so overtapped that they drop to a trickle or dry up completely for long periods of time. Water tables are falling due to the overpumping of groundwater across large areas of China, India, Pakistan, Iran, the Middle East, Mexico, and the United States. As much as 10 percent of the world's food is produced by the depletion of groundwater—a hidden water debt that creates a dangerous bubble in the food economy.[30]

It's tempting to try to solve these problems with bigger versions of familiar twentieth-century projects—especially larger dams and longer water transfers. In fact, many countries and regions are doing just that. Brazil, China, Turkey, and a number of other developing countries are on dam-building binges that make the western US experience look like a warm-up act. If completed as designed, China's $60 billion water transfer from the Yangtze River in the south to the water-short north will be the largest construction project on Earth, annually transferring a volume of water equal to half the yearly flow of the Nile River. India has an even more grandiose scheme. Called the Interlinking Rivers Project, it involves the construction of 9,000 kilometers (5,600 miles) of canals to connect thirty-seven rivers. The aim is to expand irrigation and put an end to the vicious cycles of floods and droughts that plague the South Asian nation. The estimated price tag is some $140 billion.[31]

Besides high capital costs, big engineering schemes are notorious for delays, cost overruns, and hidden social and environmental damages. Three-quarters of large dam projects end up costing nearly double the original estimate. If planners had used these actual costs in their original project analysis, many dams would be deemed economically unviable.

Similarly, the costs of mitigating social and environmental harms are either grossly underestimated or excluded altogether. China has spent $26 billion to lessen the ecological impacts of its massive Three Gorges Dam on the Yangtze River.[32]

Moreover, giant water projects often lock governments and water users into high operation and maintenance costs. For example, it requires a great deal of energy to move water, so long-distance water transfers incur steep energy bills. Southern Californians get so much of their water shipped in from the Colorado River and the northern part of the state, with some canals snaking up and over mountain ranges, that the energy used to provide household drinking water can rank third in overall household energy demands, after the air conditioner and refrigerator. Statewide, the energy required to collect, move, and treat water in California accounts for 19 percent of the state's electricity use and 30 percent of its natural gas consumption.[33]

Second, scientific understanding of "ecosystem services"—the benefits society derives from the work of nature—has advanced considerably over the last several decades. While the term *ecosystem* appeared in the scientific literature in 1935, it took another four decades for the functioning of ecosystems to be described as services to humanity. Then, in 1977, ecologist Walter E. Westman's article in the journal *Science* (provocatively titled "How Much Are Nature's Services Worth?") drove home the point that ecosystems have value.[34]

Once brought into the sphere of economics, the work of nature was on its way to becoming a necessary factor in the benefit–cost equations of development projects—whether they involved draining wetlands, building dams, or clearing forests. The Millennium Ecosystem Assessment, called for by United Nations secretary-general Kofi Annan in 2000, and which involved more than 1,300 experts from 95 countries, concluded that future human well-being depends on correcting "the historical bias that has existed against natural services when it comes to weighing the costs and benefits of particular economic choices. . . . The only 'market value' of a forest is often in the price that can be obtained for its wood, even though the standing forest may be worth much more for its contribution to water control, climate regulation, and tourism."[35]

For the last two centuries, we have been trading nature's services for engineering services. Instead of floodplains controlling floods, we built dams and levees to do that work. Instead of healthy watersheds and wetlands cleansing our water supplies, we built filtration plants to provide that service. For the most part, we viewed this substitution of technology for nature as a sign of progress. It gave society more control over water, opened up new lands for development, and spurred economic growth.

But a different view of nature gradually emerged. Natural ecosystems, when healthy and functioning well, are vital to the economy. Watersheds, wetlands, floodplains, and river systems constitute a class of infrastructure doing valuable work, just as dams, canals, and treatment plants do. Assessments led by economist Robert Costanza showed, for example, that the ability of freshwater swamps and river floodplains to store water, mitigate floods, and break down pollutants delivered annual benefits to the economy averaging some $32,000 per hectare ($13,000 per acre; both expressed in 2016 dollars). It was foolish to continue to bulldoze, dike, and drain away these services as if their value were zero.[36]

Lastly, and perhaps most importantly, the changes in weather patterns and water flows that we're beginning to see as the planet warms call into question the very assumptions that have underpinned our water projects for decades. In 2008, seven top water scientists argued persuasively in the journal *Science* that "stationarity"—the foundational concept that hydrologic systems vary and fluctuate within a known set of boundaries—is dead. When it comes to water, in other words, the past is no longer a reliable guide to the future.[37]

Take the 2012–16 drought in California, which scientists say included the worst consecutive three years of drought in 1,200 years. While lack of rain and snow was the immediate problem, unusually hot temperatures intensified the drought by 15–20 percent, according to research led by A. Park Williams at Columbia University. With the warmer atmosphere able to hold more moisture, more water evaporated from soils and surface waters, desiccating the landscape and depleting rivers, lakes, and reservoirs. Overall, the scientists calculated, California's air can now hold 32 trillion liters (8.5 trillion gallons) more water per year than it could during the cooler temperatures of a century ago. Although natural variability dominates, the Williams team concludes, global warming "has substantially increased the overall likelihood of extreme California droughts."[38]

In much of the Northern Hemisphere, climate change is also making flooding more severe. A warmer atmosphere that holds more moisture can lead to harder rains, which in turn increase the chances for disastrous flooding. Research by Myles Allen of the University of Oxford and his colleagues have linked the warming climate to the damaging floods in 2000 in England and Wales. "What has been considered a 1-in-100-years event in a stationary climate may actually occur twice as often in the future," Allen told the journal *Nature*, which published his research findings. In December 2015, the highest rainfall ever recorded in the United Kingdom occurred in the Lake District, with 341.4 millimeters (13.4 inches) falling in 24 hours. That December was both the United Kingdom's wettest and warmest on record.[39]

Similarly, scientists with the National Center for Atmospheric Research in Boulder, Colorado, published findings in 2016 that rainstorms in the United States may become more frequent and intense even at current levels of greenhouse gas emissions. The biggest effects, they found, would be in the Northeast and Gulf Coast. Intense storms could occur five times more often, and individual events could bring up to 70 percent more rain, causing more epic floods.[40] During Hurricane

Matthew in October 2016, 26 stream gauges in North Carolina and South Carolina registered record flows, according to the US Geological Survey. As residents of the historic North Carolina town of Princeville assessed the damage, many considered leaving their homes for good rather than rebuilding. After all, this was the second time they had lived through a "100-year flood" in 17 years.[41]

We are living in uncertain times. The tools that engineers use to plan investments of more than half a trillion dollars a year in dams, reservoirs, canals, and other big water projects can no longer be fully trusted. How then do we protect public health and safety, ensure food security, and manage risk? When the floods come, will the levees hold? With more severe droughts likely, will the reservoirs refill? Does it make sense to build a big new dam if the water it holds back may be insufficient to generate hydroelectric power? Will massive amounts of sediment eroded from mountainsides by intense rainstorms fill a new reservoir with sediment and cut short its useful life? Will farms get the irrigation water they need once the glacier-fed river flows have dwindled? How do we plan for what once seemed unthinkable—the disappearance of prime water sources for cities, industries, and farms? In short, how do we live with these new realities?

Decades ago, Albert Einstein reminded us of a fundamental lesson that's hard to learn: "We can't solve problems by using the same kind of thinking we used when we created them." Fortunately, just when it's crucially needed, a new mind-set about water is taking shape. It's one that blends engineering, ecology, economics, and related fields into a more holistic approach that recognizes the fundamental value of nature's services.

As the chapters that follow will show, this evolving mind-set is already changing the way we manage water. Working with, rather than against, nature, pioneering cities, farmers, businesses, and conservationists are rejuvenating watersheds and floodplains, and replenishing rivers, groundwater, and soils. The result is a smarter way to mitigate flood damages, prepare for droughts, restore habitats, grow food, augment water supplies, and generally strengthen water security. Investing in a healthier water cycle, it turns out, may be the best insurance policy money can buy in this century of rapid change.

CHAPTER 2

Back to Life

A faint bugle note soon told us they were cranes, inspecting their
Delta and finding it good.
—*Aldo Leopold*

ON A SUNNY TUESDAY AFTERNOON in the bustling border town of
San Luis Rio Colorado, Sonora, Mexico, word got out that the river
was coming. Local residents streamed in from all directions and gathered beneath the San Luis Bridge, which connects the Mexican states of
Sonora and Baja California. Families picnicked to pass the time. Dogs
merrily chased balls through the sand. And in one last showy display
before the water arrived, young men spun their pickup trucks around
the dry river channel. A certain buzz filled the air. The kids growing up
here had never even seen the river that had given their town part of its
name. But in just a couple of hours, the Colorado River would arrive.

Two days earlier, on March 23, 2014, engineers had opened the gates
of Morelos Dam, the last of the long line of dams that block the flow
of the Colorado River as it journeys 2,330 kilometers (1,450 miles)

from the US Rocky Mountains to Mexico's Sea of Cortez. Ever since the completion of Glen Canyon Dam in 1963, the Colorado, with a flow powerful enough to sculpt the Grand Canyon, had rarely passed San Luis and coursed through its delta to the sea. It gave its water up to Los Angeles and San Diego, Phoenix and Tucson, Las Vegas, Denver, and Mexicali. It made the deserts of the Southwest burst with lettuce, tomatoes, melons, alfalfa, and cotton.

The Colorado River Delta was once a lush, watery paradise spanning some 8,000 square kilometers (3,000 square miles) and teeming with quail, deer, bobcat, jaguar, and vast fleets of waterfowl. Its wetlands and cottonwood–willow forests once provided a crucial stopover for millions of migratory birds on the Pacific Flyway. But deprived of water for most of the last half century, the delta had become a desiccated place of salt flats, dry channels, and invasive salt cedar. There were few places for the avian multitudes to rest, feed, and breed.

Fisherfolk in the upper Sea of Cortez (also known as the Gulf of California) suffered from the river's disappearance too. For millennia, the Colorado River's nutrient-rich freshwater mixed with the upper gulf's salty tides to create the perfect water chemistry and nursery grounds for gulf corvina, totoaba (a cousin of white sea bass), brown and blue shrimp, and other valuable fisheries. But without the river's freshwater flowing in, the estuary became too salty. The indigenous Cucapá, the "people of the river," had fished and farmed in the delta for at least a thousand years, keying their lives to the river's ebb and flow. But without the river, they were a dying culture. Today only some 200 Cucapá live in the delta region.[1]

How these changes happened is no big mystery. In November 1922, representatives of all seven US states with land in the Colorado River's watershed traveled to Bishop's Lodge outside of Santa Fe, New Mexico. There they were met by then secretary of commerce Herbert Hoover, whose job it was to broker a deal to divvy up the liquid lifeline of the

American Southwest. But around that table, several important voices were missing, including those of Mexico and the river itself. A treaty between the two countries signed in 1944 at least partially corrected the first of those omissions. It allotted 10 percent of the river's flow to Mexico and 90 percent to the United States. However, an allocation to sustain the delta's wetlands, trees, fisheries, birds, communities, and cultures didn't factor into the discussion.

Meanwhile, engineers and laborers had already gotten to work building the infrastructure needed to deliver water and power to the seven US states that had signed the 1922 compact. Completed in 1935, Boulder Dam (later renamed Hoover Dam) created Lake Mead, still the largest man-made reservoir in the United States. Davis, Imperial, Parker, and a half-dozen other dams followed. With each reservoir to store water and each canal to deliver it to the burgeoning cities and farms throughout the basin, the Colorado became less like a river and more like an elegant plumbing system. It was no longer sculpting nature's canyons and terrain so much as fashioning human landscapes—from irrigated farm fields as far as the eye could see to rapidly expanding urban oases in the desert.

After signing the 1944 treaty, Mexico built one dam on the lower Colorado—Morelos, named after a patriotic leader of its independence movement. The dam is situated 2 kilometers (1.2 miles) downstream from where the boundary of California and Baja California intersects the river. Morelos was designed only to divert water, not to store it. At the dam, Mexico's 10 percent share of the river flows into a big canal, Canal Central, which delivers water to the Mexicali Valley. Because the compacts and laws that divided up the river promised more water to the seven US states and Mexico than the river typically carries, in most years over the last half century the Colorado has been drained completely dry at Morelos. Below the dam it becomes, in effect, a river no more.

So as the gates of Morelos opened on that Sunday morning in March

2014, cheers rang out from those who had gathered to witness a most improbable event: the intentional release of water from an overallocated, drought-stricken, binational river for the purpose of mimicking what the river had done naturally for millennia—send a pulse of floodwater through its delta in the spring. As with most rivers fed by snowmelt, the Colorado's spring pulse is crucial to its natural flow. The flood cleanses the channel, helps cottonwoods and willows grow along the riverbanks, replenishes backwaters and floodplain wetlands, and delivers nutrient-rich sediment and freshwater to the estuary. Without that pulse of water, the habitats of the delta—and all the life and economic activity they supported—could not thrive.

I first traveled to the Colorado River Delta more than two decades ago, in the spring of 1996. At that time, the common narrative, at least in the United States, was that the delta was dead. It was certainly forgotten; some US maps even showed the river ending at the international border. For the Colorado, it was an unfortunate circumstance of political geography that its last 153 kilometers (95 miles) were in another country. The river had become the lifeline of the American Southwest, but on the US side of the border its crucial role for Mexico's land and people barely registered.

Much of what I saw during that trip in 1996 did indeed fit the narrative of a desiccated wasteland. It was hard to believe such a place could ever come back to life—except, much to my surprise, a portion of it already had. As I flew over the delta in a small plane with University of Arizona biologist Edward Glenn, an oasis of green suddenly appeared in the barren brown landscape. It was a wetland called the Ciénega de Santa Clara, a maze of lagoons ringed by cattails and marsh grasses. It was an astounding sight—and it had formed completely by accident.

On an ordinary day in 1977, a young man named Juan Butrón was out walking not far from his home in the small delta town of Ejido

The Colorado River Delta.

Johnson. He stumbled upon a large canal filled with water and followed it to its end, where he was astonished to find a shallow lake spread out before him. Returning regularly, he saw fish appear. Then cattails, reeds, and bulrushes sprang up, turning the lake into a web of marshes and lagoons. Butrón began to picnic and fish there with his family and neighbors. Eventually he and other Ejido Johnson residents started offering "eco-tours" of the marshy wonderland. Over the years the Ciénega would expand and shrink along with the volume of drainage flowing in, but the marsh now covers more than 6,100 hectares (15,100 acres), and the mudflats just to the south span roughly an additional 10,000 hectares (24,700 acres).[2]

It is no exaggeration to say that today the Ciénega de Santa Clara is one of the most important desert wetlands in North America. Yet for years it was so little known to the outside world that it did not receive an official place-name until 1992. Today the Ciénega supports 280 species of birds, including the elusive and endangered Yuma clapper rail. Thousands of migratory birds stop to rest, feed, or spend their winters there.[3]

The canal Butrón discovered originates in the Wellton-Mohawk Irrigation and Drainage District, a farming region in southern Arizona. As irrigation water seeps through farmland soils, it often picks up salts, pesticides, and other chemicals that make the drainage coming off the fields highly polluted. Back in the 1960s, Mexican officials justifiably complained to their northern neighbor that its delivery of this salty drainage to Mexico as part of its Colorado River allotment was killing farmers' crops across the border. The US Bureau of Reclamation responded by building a concrete-lined canal roughly 80 kilometers (50 miles) long from the irrigation district in southern Arizona into the eastern part of the delta in Sonora, Mexico. The canal, known as the MODE, was capable of transporting 130 million cubic meters (105,000 acre-feet) of the district's agricultural drainage into the delta each year. In effect, the United States used the delta as a dumping ground for a volume of agricultural wastewater equivalent to 0.7 percent of the Colorado's historical annual flow.

The releases to the delta began in 1977, not long before Butrón first came upon the canal, and they were intended to be temporary. The US plan was to treat the salty drainage in a new desalination facility in Yuma, Arizona, and then use the water again. But the desalting plant never fully operated. Meanwhile, scientific research by Glenn, who first visited the Ciénega in 1979, and others shined a light on the ecological importance of this wetland and the fact that it would disappear if the drainage water went instead to the Yuma desalting plant. In 1993, the Mexican government declared the upper Gulf of California and the Colorado Delta, including part of the Ciénega, an international biosphere reserve, a designation sanctioned by the United Nations and intended to protect world-class ecosystems.

The lesson of the Ciénega was plain and simple: if we just add water, in this case even low-quality water, habitats can come back. Parts of the delta could live again. That's what Glenn was showing me on that flyover in 1996. When I returned in February 2013, I traveled again to the Ciénega, this time with a former graduate student of Glenn's named Osvel Hinojosa-Huerta, director of the Water and Wetlands Program for Pronatura Noroeste, the regional chapter of Mexico's largest conservation organization. He amplified Glenn's message. "It is so resilient," said Hinojosa-Huerta, who has been studying the ecosystems of the Colorado Delta since 1998. "Life just wants to return here."

Along with Hinojosa-Huerta and Butrón, who by now knew the ins and outs of this wetland perhaps better than anyone, I climbed into a boat and plied the marshes. American coots glided smoothly along the water's surface. A small riot of bird sounds erupted from the cattails. Out on the sandbars, black-neck stilts did their circus walks and long-billed dowitchers poked deep into the mud for late afternoon snacks. Then, from some distant corner of cattails came a "keck, keck, keck, keck." Hinojosa-Huerta turned to me and whispered, "A clapper rail."

Hinojosa-Huerta is an expert on the Yuma clapper rail, now an endangered bird due to the loss of its habitat. Every year since 1998 he has organized surveys to assess Mexico's clapper rail population. He estimates that some 6,000–7,000 individuals live in the Ciénega, about three-quarters of the remaining global population.

As the setting sun turned the desert sky into a quilt of brilliant reds and oranges, a pair of northern shovelers flew by. Huge flocks of swallows swooped through the air. I recalled what the naturalist Aldo Leopold had experienced when he canoed through the delta with his brother Carl in 1922, ironically the same year the Colorado was divvied up at Bishop's Lodge. A "milk and honey wilderness," Leopold called the delta. A land of "a hundred green lagoons." The river that for Leopold was "nowhere and everywhere" as it slowly meandered its way toward the sea was no more. But here, an accidental wetland sustained by salty farm drainage that could be cut off in a flash served as a precious reminder of the delta's former glory and a beacon of hope for its revival.[4]

~

In 1995, the year before I first traveled to the Colorado Delta, I was asked by the World Bank to join a fact-finding mission to another unique ecosystem that had lost its liquid sustenance—the Aral Sea in Central Asia. Once the world's fourth-largest lake, the Aral had been shrinking for several decades because Soviet leaders in Moscow had calculated that the two rivers flowing into the sea, the Amu Darya and Syr Darya, would be more valuable if diverted to grow cotton in the desert. Once the size of Ireland, the Aral was evaporating into the desert air. So in March of that year, I spent a short time in Uzbekistan and Turkmenistan with an international team of scientists and World Bank staff. Despite all the preparation I'd done before the trip, what I saw and experienced there shocked me.

I gazed out from a bluff on the outskirts of Muynak, a former seaside port town, but I saw no water—just an endless expanse of desiccated earth and a graveyard of ships in the dried-up seabed. The Aral had already split into two, a small northern sea and a much larger southern one. The larger sea had also begun to split into western and eastern lobes. I learned that on windy days, toxic dust storms made the air hazardous to breathe. The lake's fish had died off, and with them went 60,000 fishing jobs. Muynak had the feeling of a ghost town, as thousands of "ecological refugees" had fled the area. The people who remained suffered from startlingly high rates of anemia, respiratory ailments, and cancers. Babies and infants died in high numbers. Never before had I grasped so viscerally the connections between the decline of an ecosystem and the decline of the people who depend on that ecosystem.

At a meeting in Nukus, the capital of an autonomous region of Uzbekistan called Karakalpakia, our team spoke with local health specialists, environmentalists, and community groups striving to make life better in this disaster zone. They spoke of illness, degradation, death, and deep sadness over the loss of their beloved Aral Sea and way of life. I listened. And then they asked us, the so-called experts from the other side of the globe, what could be done. I felt almost paralyzed. Given the political and economic realities of the time, I saw no honest hope for a revival of the sea and the communities that had lived around it.[5]

The Aral Sea did indeed continue to shrink. By 2005, a decade after I was there, the lake had lost 80 percent of its water. Satellite imagery released by the European Space Agency in 2009 showed that the eastern lobe had lost an additional 80 percent in just the previous three years. And then, in August 2014, an image from NASA's Terra satellite just about knocked my socks off: it revealed that the eastern lobe had completely dried up. Except for the small northern lake, which has been blocked off in a desperate attempt to save a remnant of the original water body, the Aral may disappear in my lifetime.

Around the time of my back-to-back visits to the Aral Sea and the Colorado Delta in the midnineties, a larger picture of what was happening to the world's rivers had started to take shape in my mind. The hydrologic data back then weren't nearly as good as they are now, but I examined as best I could trends in river flows from around the world. It turned out that it wasn't just the Amu Darya and the Colorado that were running out for months or years at a time, but many other rivers as well—including the Indus and the Ganges in South Asia, the Nile in northeastern Africa, the Murray in Australia, and the Yellow in China. These were major rivers, collectively supplying hundreds of millions of people and millions of hectares of irrigated land. In 1995 I wrote an article for the Worldwatch Institute, where I had previously worked, titled "Where Have All the Rivers Gone?"

Although the details varied, the answer to this question was similar in each case: dams and diversions had siphoned away the river's flow to supply water to expanding cities and farming regions. As with the Colorado agreement signed at Bishop's Lodge, little thought had been given to the people and ecosystems at the end of the line. The result was the decline of river deltas, coastal fisheries, and aquatic diversity—as well as the cultures and economic activities tied to these river and estuarine ecosystems.

The argument went, and often still goes, that this is a necessary and smart trade-off. After all, populations are growing, and people need food, energy, and drinking water. In order to thrive, an economy must have water and power. The jobs and income created by putting water to use on farms and in factories are worth the loss of wetlands, wildlife habitats, fisheries, and even cultures and livelihoods downstream.

Except, I was discovering, this argument is a red herring. Water is so heavily subsidized and delivered at such low costs—especially to farms, which consume the vast majority of water in areas of shortage—that it was not being used anywhere near as efficiently and productively as possible. Water laws grounded in the "use-it-or-lose-it" principle encouraged farmers to use more water than they really needed so as to avoid losing their valuable water rights, and they had limited, if any, ways to sell water they didn't really need. It turns out there's enough water to maintain at least some of the functions and economic benefits of healthy rivers and wetlands if we get smarter about how we manage water.

～

I doubt that the kids of San Luis Rio Colorado would have been waiting expectantly for the return of their city's namesake river on that Wednesday afternoon in March 2014 if an accidental wetland created by farm drainage had been the only sign of the delta's possible resurrection. But the Ciénega de Santa Clara wasn't the only sign of hope. Thanks to water releases of another kind, even the great cottonwood–willow forests so crucial to the migratory birds, wildlife, and natural beauty of the delta had shown an ability to revitalize.

Since the early 1980s, spring floods have inundated the Colorado Delta when two conditions coincide: the giant reservoirs behind Hoover and Glen Canyon Dams are full, and the upper Colorado watershed receives a large amount of precipitation. The latter often happens with the arrival of the weather phenomenon known as El Niño. In late 1982, for example, Lake Powell had just finished filling up for the first time after the completion of Glen Canyon Dam. No water had reached the delta for the prior 18 years. Then an El Niño cycle began. A few months later, in the spring, a huge amount of snowmelt and rainwater rushed into the upper Colorado River. With both Powell and Mead filled to capacity, engineers passed the floodwaters through the reservoirs and sent them downstream, where they then flowed through the gates at Morelos Dam and on through the delta. Similar conditions persisted for several years, such that from 1983 through 1986 the total flow into the delta averaged 12.8 billion cubic meters (10.4 million acre-feet)— the biggest discharge in a four-year period since the 1920s, before any big dams blocked the river. Smaller but still substantial floods coursed through the delta in 1993, 1997, 1998, and 1999.[6]

As in the Ciénega, the delta awakened. Satellite images taken not long after the mideighties' floods showed that about 40,000 hectares (100,000 acres) of habitat along the river's channel had sprung back—a ribbon of green through a brown landscape. The floods had helped cottonwood seeds to germinate and spread. But as the flood flows diminished and then disappeared, so did the riverside forest. After the last significant El Niño flood in 1999, the delta's riparian habitat shrank back to about 2,600 hectares (6,500 acres).[7] Even at this size, though, it was far larger and better habitat for birds and wildlife than existed anywhere along the river on the US side of the border, and it revealed the recipe for the delta's revival: just add water.

But where would that key ingredient come from? Neither the agricultural drainage water sustaining the Ciénega nor the floodwaters that greened-up the delta's riparian areas were at all secure. Starting in 2000, an unrelenting drought gripped the Colorado River basin. The next 14 years would turn out to be the driest the basin had seen in a century. Demand for the river's water exceeded the sustainable supply. Lakes Mead and Powell, just about full in 2000, shrank to half their combined capacity. Researchers at the Scripps Institution of Oceanography at the University of California–San Diego reported there was a 50 percent chance that Lake Mead could effectively be dry by 2021. Clearly no surplus flows would be coursing through the delta anytime soon.[8]

But while Glenn and other scientists were documenting the resurgent biology of the Colorado Delta, a group of economists, policy analysts, and conservationists began exploring options for allocating some water to the delta. In June 1999, the Environmental Defense Fund (EDF) published a report titled *A Delta Once More: Restoring Riparian and Wetland Habitat in the Colorado River Delta*. In it, the binational team of authors recommended that US and Mexican officials begin negotiations to amend the 1944 treaty so as to provide an allocation of water for the delta. They recommended the delivery of a flood or "pulse" flow once every four years, along with perennial or "base" flows to sustain 60,000 hectares (150,000 acres) of wetland and riparian habitat. The total volume of water recommended for the delta per year amounted to less than 1 percent of the Colorado River's total annual flow.[9]

It was a practical strategy built on a sense of the possible. The conservationists did not call for dismantling big dams or drying up irrigated agriculture. Instead, they made a compelling case to provide a very modest volume of water to the delta that scientific studies showed, if delivered at the right time and to the right places, could accomplish a great deal of ecological good.

The strategy worked. These recommendations, followed by a 2005 report laying out the conservation priorities of the delta in more detail, laid the foundation for what would become a landmark agreement between the United States and Mexico to return some water to the delta.[10] In November 2012, after years of binational scientific collaboration, political negotiations, strategic deal making, and no small measure of patience and persistence by all the parties involved, the two countries signed an addendum called Minute 319 to their 1944 treaty. It began a five-year experiment to return water to the delta in the form of a pulse flow and sustaining base flows.[11]

With the whole Colorado basin still in drought and the levels of Lakes Mead and Powell continuing to drop, the signing of Minute 319 was nothing short of miraculous. "Leaving behind unilateralism, the two countries united to sign the most important bilateral Colorado River agreement since the 1944 Treaty," declared Jennifer Pitt, then director of EDF's Colorado Delta program and coauthor of both the 1999 and the 2005 stage-setting reports. She had worked tirelessly for some 14 years to make this historic event happen. Along with Pitt, the core members of the conservation coalition that helped bring the deal to fruition included Arizona-based water rights attorney Peter Culp, ecologist Hinojosa-Huerta with Pronatura Noroeste, delta water market expert Yamilett Carrillo, and resource geographer Francisco Zamora with the Tucson-based Sonoran Institute.

Much of the genius behind Minute 319 was in figuring out how to give all sides something they really wanted. The agreement allows Mexico to store water in Lake Mead, which gives both nations more flexibility in how they manage the Colorado's water. It sets out a new formula for sharing both the benefits of "surplus" water and the pain of shortages. As for the conservation community, they got the prize they'd been striving toward for some dozen years: water dedicated to the revival of the delta.

The ink was barely dry on the document when the binational scientific and conservation teams got busy preparing to execute the grand experiment. Minute 319 called for a five-year pilot project to provide a total of 195 million cubic meters (158,088 acre-feet) of water to the lower river and its delta. That's about 1 percent of the river's historical annual flow delivered over five years. Although Glenn and others had recommended that a similar volume be delivered on average *every* year, scientists were still confident the water could yield significant ecological benefit if strategically timed and delivered.

Two-thirds of the Minute 319 water would be used to deliver a pulse flow designed to mimic the natural flood that historically had occurred every spring. The two governments had equal responsibility for coming up with the water for the pulse. A key principle Pitt and others had put forward prior to the negotiations of Minute 319 was that of "conservation before shortage," which essentially meant that the two governments should work together to conserve water and store it in Lake Mead in order to avoid the declaration of a shortage, which the US government was obligated to do if Mead's level dropped below 328 meters (1,075 feet) above mean sea level. So by working together to conserve some of Mexico's water and store it in Lake Mead, the two governments could reduce the likelihood of a shortage declaration (and resulting water supply cutbacks) and make water available for a pulse flow.

With a few modifications, this is what the two countries did. Because of damage to some of its irrigation infrastructure in an April 2010 earthquake, Mexico was unable to use its full water allotment, and so it ended up storing some "surplus" water in Lake Mead. For its part, the United

States promised to help line irrigation canals in the Mexicali Valley and financially support other conservation measures in order to come up with its share of the pulse water, even if retroactively.

The conservation coalition had responsibility for coming up with the roughly 65 million cubic meters (52,696 acre-feet) needed for the base flows over the five years of the pilot program. Without those sustaining flows, the trees and vegetation in newly created habitats would die. To acquire water for these base flows, the coalition formed the Colorado River Delta Water Trust, the first water bank in Mexico aimed at returning water to the environment. The Trust buys and leases water from willing farmers in the Mexicali Valley, and then works with Mexican water authorities to deliver that water through existing irrigation infrastructure to the newly created delta habitats. Funding for the Trust's water acquisitions comes from an array of donors, foundations, and conservation organizations on both sides of the border.

Fortunately, a very active water market already existed in the Mexicali Valley prior to the Minute 319 restoration, so the region's roughly 12,000 farmers were accustomed to the idea of buying and selling water rights. Most have small farms of less than 20 hectares (50 acres). About 70 percent of the land is planted in wheat, while the remainder is in cotton, alfalfa, or vegetables. At the helm of the Delta Water Trust was Yamilett Carrillo, then a consultant to Pronatura Noroeste who had studied the intricacies of the region's water market as part of her doctoral work at the University of Arizona in Tucson. She is now executive director of the independent nonprofit organization into which the Delta Water Trust evolved, Restauremos El Colorado AC.

Farmers in the Mexicali Valley sell their water rights for different reasons, Carrillo told me during my visit to the delta in February 2013. Some wish to retire and do not have children who want to take over the business. Others pass away, and their families wish to put the water rights up for sale. In some cases it's purely good economics. After the

2010 earthquake and resulting damage to the region's irrigation infra-structure, the value of water rights roughly doubled. Some wheat farmers found their water rights to be worth seven times their annual income. Moreover, many farmers favor selling their rights to the Delta Water Trust rather than to Tijuana because they prefer that the water be used locally to revitalize the landscape, create jobs, and bring back the trees and songbirds of bygone days.[12]

Finally, after intense months of preparation and design work by teams of scientists, the big day arrived. On the morning of Sunday, March 23, 2014, engineers with the International Boundary and Water Commission lifted the gates of Morelos Dam. EDF's Jennifer Pitt popped champagne and sprayed it over Carrillo, Culp, Hinojosa-Huerta, and Zamora. It had been a long, crazy ride, but they had done it.

"For so many years it felt like a little band of dreamers," Pitt later told me in her office in Boulder, Colorado. "The idea that we could actually see it happen was incredible."[13]

Four days after the Morelos gates opened, a dozen dignitaries from Mexico and the United States faced a crowd of more than 200 gathered at the dam to officially celebrate the momentous achievement of Minute 319.

"We are witnesses to history," proclaimed Michael Conner, deputy secretary of the US Department of the Interior. Conner lauded the "extraordinary work of the NGO [nongovernmental organization] community on both sides of the border" for its role in making the historic event possible. Just a few years ago, Conner noted, experts would have said this landmark achievement could never happen.

For me, watching the Colorado River reclaim its channel, rediscover its delta, and flow toward the sea was one of the highlights of my professional life. The freshwater restoration program called Change the Course that I helped create during my work with the National Geographic Society had committed funds to the Delta Water Trust to purchase water for the base flows. I had gone to the delta with a team from

National Geographic to video, photograph, research, and write about this historic event.

Just a few days after the flow release began, Hinojosa-Huerta and I hopped into canoes just downstream of Morelos Dam and paddled the upper reaches of the Colorado River. Juan Butrón, who by that time worked with Hinojosa-Huerta at Pronatura Noroeste, was with us too. The river was flowing full, spreading onto its floodplain, and nourishing native cottonwoods and willows that hadn't gotten a good drink in quite some time. We watched as the wind sent their seeds aloft. Many landed on the water's surface to hitch a ride downstream. Within days, the seeds would nestle into a moist bank to germinate. Scientists had done their best to time the pulse flow with the window of seed germination, and the hope was for thousands of new cottonwoods and willows to spring up as a result. Of course the nonnative salt cedar (also known as tamarisk) would benefit from the pulse of water too.

Ecologist Osvel Hinojosa-Huerta surveys birds in the upper reaches of the Colorado River Delta. Photo by Cheryl Zook/National Geographic.

As we paddled downriver, we eyed three muskrats along the bank. Ducks flew overhead, while rails and bitterns appeared to be looking for places to nest. It was migration time for warblers, sparrows, and thrushes, and they were following the corridor of trees alongside the river. "They react to cues," said Hinojosa-Huerta, who can identify about 350 bird species just by their calls. "They see green and structure, and know it's good."

While canoeing 15 kilometers (10 miles), we saw or heard 40 different species of birds—including white-tailed kite, ash-throated flycatcher, belted kingfisher, and four varieties of heron. It was music to an ecologist's ears—and a sign of water bringing life to the delta.

For the duration of the pulse flow, teams of scientists fanned out across the delta to document the results of this unique experiment. Jorge Ramírez-Hernández, a hydrology professor at the Universidad Autónoma de Baja California, supervised two teams of students tasked with measuring the Colorado's flow every day at 10 different sites. They used a method called Acoustic Doppler Current Profiler, which applies principles of sound propagation through water to measure flow velocity. When combined with the channel's cross-sectional area, it provides an estimate of the river's discharge. Other students were monitoring groundwater levels to detect the influence of the pulse on the underground water table. When we joined Eliana Rodriguez Burgueño, she was 300 meters (984 feet) from the still-dry main channel dropping a piezometer into a monitoring well, a test she'd been doing monthly for six years. To have her dissertation research coincide with such a historic event was a big stroke of luck. "This is an experience of a lifetime," she said.

Just as we were heading out, Ramírez got a phone call. The river would reach San Luis Rio Colorado by mid- to late afternoon. We hopped in our Jeep and headed to the San Luis Bridge.

Once there, my National Geographic colleagues set up time-lapse cameras while I walked up the broad, sandy channel. The locals who'd gathered were waiting expectantly. Then, around three o'clock, a shimmering mirage appeared up the channel. It had taken about 55 hours to get here from Morelos Dam, but the river had arrived. It advanced slowly, but before long, the sands of the channel became moist, then wet, and then suddenly we had to move quickly to keep our boots from getting soaked. We watched the channel fill not only with water, but with members of the community. Little Isabella Cedillo Castro, wearing a bright yellow shirt and matching ponytail tie, kneeled in the water and raced against time to complete her sand castle before the rising river overtook it. Two women strolled alongside the flowing water beneath a bright red sun umbrella. In whatever way felt right, the residents of San Luis Rio Colorado welcomed their river home.

When we returned to the bridge on Sunday afternoon, a week after the gates of Morelos were opened, the river was flowing deep and strong.

Hundreds of people lined the riverbanks near the bridge to celebrate. Music and barbecue smells filled the air. The atmosphere was like a carnival. Victor Reyes Cervantes, a late-middle-aged man who sells tractors for John Deere, recalled the river flowing this big back in the 1950s. He said he hoped that the fish, birds, and wildlife would come back along with the river. A young mother keeping watch over her two-year-old son, Leonardo, as he played near the water's edge, nodded in his direction and talked of how sad he'll feel if the river goes away. Roxana Torres, a college student, worried about the trash the river was accumulating and the disappointment that lay ahead for the community when the pulse flow ended. "I believe," she said, "that people think the river will stay from now on."

On the eighth day after the initial release from Morelos, I witnessed something extraordinary. As on most mornings, I headed out with my colleagues before dawn to find the leading edge of the river as it made its

Children of San Luis Rio Colorado play in their town's namesake river during the historic pulse flow of 2014. Photo by Cheryl Zook/National Geographic.

way toward the sea. That morning, as often happened during those days tracking the river, I ran into a group of scientists who were studying this unprecedented ecological experiment.

Along with Karl Flessa, professor of geosciences at the University of Arizona in Tucson and co-chief scientist of the monitoring team for the pulse flow, were freshwater biologist Rebecca Lester and marine ecologist Jan Barton from Deakin University in Victoria, Australia. Lester and Barton were colleagues of Flessa's, and they had come all the way from Australia to witness this grand experiment firsthand. They were crouched in the channel and staring intently into the dark green rim of the river's edge as it inched along. "Copepods," they said excitedly as I kneeled beside them.

Sure enough, the rim of the river was alive. These microscopic crustaceans had lain dormant in the desert sands for a decade or more. The

Conservationists, including the author (on the left, with notebook), observe the Colorado River as it slowly reclaims its channel in the delta and flows toward the sea. Photo by Cheryl Zook/National Geographic.

females lay a kind of leathery egg that can remain viable for many years, even through extreme dryness. Within a couple days of being wetted by the river's flow, billions of tiny copepods had hatched. Some were now feeding on algae along the river's fringe.

Just as I was taking in this mini miracle, Lester exclaimed, "Dragonflies are coming!" And sure enough, scooting along the river, drawn to the water to breed, were these big-eyed, winged insects. Dragonflies eat copepods, and they were on the hunt. Then came carp, which eat dragonflies. Lester had also seen fish larvae eating the copepods. "This is exciting," she said. "You can see the food web developing within minutes of the water arriving." It was the most literal experience one could imagine of the "water is life" maxim.

～

On May 15, after coursing through its delta for nearly eight weeks, the freshwaters of the Colorado River touched the salty tides of the upper gulf. If rivers are born with a destiny, it is to reach the sea. For that brief moment in time, two nations and a team of dedicated scientists and conservationists had enabled the Colorado to reach hers.

After the drama of the pulse flow was over, media interest largely faded away, but the hard work of restoration was just gearing up. The ultimate goal is to create what Hinojosa-Huerta calls "stepping stones of habitat" that allow birds and wildlife to find enough places across the delta to feed, breed, and rest. One of those sites, Miguel Alemán, is located in an old meander of the Colorado River. There, Pronatura Noroeste has overseen the removal of invasive salt cedar and the planting of thousands of native cottonwoods, willows, and mesquites, which were grown from seed in a nearby nursery. A pipe carries water from an irrigation canal over to the site to irrigate the seedlings. Trees planted during the spring of 2014 grew 3 meters (10 feet) by the end of the year.[14]

Farther downstream at Laguna Grande, the Sonoran Institute has overseen the planting of many thousands of cottonwoods, willows, and mesquites. This area is beginning to resemble what the lower delta looked like back when Aldo Leopold canoed through it in 1922. Ultimately, this site, which has the advantage of relatively shallow groundwater to support the trees' growth, will encompass some 445 hectares (1,100 acres). In October 2016, Sonoran held its fifth annual tree-planting event, bringing volunteers together with nonprofit groups and government agencies to expand the habitat at Laguna Grande. The community park at the restoration site allows the public to spot resident and migratory birds, as well as wildlife drawn to this desert oasis, including bobcats, coyotes, and beavers.[15]

Following the pulse flow, which raised the water table, spurred native trees to germinate, and nourished the actively planted trees at Laguna

Grande and other restoration sites, Restauremos (the water trust) is working with Mexican water authorities to deliver base flows to support the trees' growth. By the end of 2016, the Trust had delivered 61.6 million cubic meters (49,979 acre-feet) of base flows to the delta restoration sites, 95 percent of the volume required under Minute 319.[16]

The scientific teams have documented numerous benefits of the pulse and base flows delivered to the delta—including the beneficial recharge of groundwater, an increased number of migratory birds in open-water areas, and an overall green-up of the landscape. In their monitoring report released in October 2016, which covers results through early December 2015, the scientists reported that the pulse flow had recharged groundwater and flushed salts from the soil, both of which aid the growth of native willows, cottonwoods, and other delta vegetation. They found that between 2013 and 2015 the combined abundance of 19 bird species of conservation interest—including the Gila woodpecker, the ash-throated flycatcher, and the yellow-breasted chat—had increased 49 percent. Moreover, those 19 species were 43 percent more abundant at the restoration sites than in the rest of the floodplain, highlighting the habitat values of the active restoration areas. The scientific team was not surprised to find that the pulse flow had little or no impact on fisheries in the upper gulf, given the small volume of Colorado River water that actually reached the sea.[17]

"Some of the cottonwoods that germinated during the initial pulse flow are now more than 10 feet tall," said Karl Flessa, the University of Arizona professor serving as co-chief scientist of the Minute 319 monitoring team, upon release of the assessment. "This short-term event has had lasting consequences. This really demonstrates that a little bit of water does a lot of environmental good."[18]

Through their on-the-ground measurements and monitoring, scientists are getting a good idea of what worked well and what to do differently if a pulse flow is again delivered to the delta. (A third and final report will be prepared after the Minute 319 experiment ends in late 2017.) As of early 2017, negotiations for a successor to Minute 319 were still under way. Carrillo, the head of the new water trust for the delta, Restauremos, hopes her organization will be able to acquire enough water to allow a couple of gates at Morelos Dam to remain open year-round, benefiting habitats in the upper reach, and to deliver enough water through irrigation canals to keep stretches of the river farther downstream, near the Laguna Grande restoration site, flowing continuously. "Someone once told me," Carrillo wrote to me in early 2017, "that 'managing water is managing conflict.' But Restauremos has shown me that managing water can also be managing cooperatively to get a river flowing again."[19]

The revival of the Colorado River Delta ranks among the most remarkable—and improbable—conservation triumphs in modern times. In many ways, it gives *resilience* a whole new meaning. And yet these gains are tenuous; without a sustained commitment of water, they will evaporate. It was the coming together of two governments, a binational team of scientists and conservationists, the support of foundations and donors, and a small dose of serendipity that made the delta's revival possible. How much life returns, and ultimately lives on, is now largely in human hands.

CHAPTER 3

Put Watersheds to Work

Eventually, all things merge into one, and a river runs through it.
—*Norman Maclean*

It's hard to believe that any city in Brazil, the South American nation sometimes called the "Saudi Arabia of water," would generate a new class of water refugees. But after two consecutive years of a punishing drought, some residents of São Paulo—a megacity of 20 million people and Brazil's economic heartbeat—decided to leave. Some had gone without tap water for days at a time; many endured half-day stints of dry taps. Resorts next to dried-out reservoirs were forced to lay off staff and shut their doors. Economists warned that the drought could shave as much as 2 percent from the nation's GDP.[1]

São Paulo is no desert city. It averages 1,455 millimeters (57.3 inches) of rain per year, nearly on par with New Orleans. But in 2014–15, the worst drought in eight decades gripped the megalopolis. The five reservoirs in the Cantareira system, which normally provides nearly half of the city's drinking water, dropped to 5 percent of capacity. Poor management of the city's water system, including leaky pipes that lost 30 percent of the water supply, made the drought even worse. But scientists also pointed to another factor with potentially severe and long-term consequences: the deforestation of the Amazon rainforest and resulting disruption of the region's water cycle.[2]

The cycling of water across the Amazon begins with easterly trade winds transferring moisture from the Atlantic Ocean to the land, where it falls as rain. Trees then capture at least half of this rainfall, whether by intercepting it directly or by pulling it up from the soil through their roots. This moisture then returns to the atmosphere as vapor as it evaporates or transpires through the leaves. There it condenses into clouds, moves westward with the winds, and falls again as rain. This continuous recycling of moisture westward across the Amazon basin forms a "flying river" that turns southward when it hits the vertical barrier of the Andes Mountains and then curves back east, bringing rains to southeastern Brazil, including São Paulo. It's an elegant system and one of the best examples in the world of how wind, rain, sun, soils, and trees work together to create regional climatic conditions and deliver reliable water supplies.[3]

More than three decades ago, researchers Eneas Salati and Peter Vose of the University of São Paulo described this unique system of moisture delivery in the Amazon basin and warned of the dangers of massive tree cutting. "Continued large-scale deforestation," they wrote in the journal *Science*, "is likely to lead to . . . reduced evapotranspiration and ultimately reduced precipitation."[4]

About one-fifth of the Amazon basin, some 1.4 million square kilometers (540 thousand square miles), has been cleared to make way for crops and cattle. Another one-fifth or so is heavily degraded. The massive road networks being built to pave the way for the construction of hundreds of hydroelectric dams are opening vast new areas of the basin to logging and forest clearing. In recent years, scientists have begun sounding alarms about the impact of deforestation combined with climate change. As the region dries out and more fires burn, the already diminished Amazon forest ecosystem could tip into a different state.[5]

For São Paulo there is little comfort to be found in these trends. The conveyor belt of moisture that crosses the Amazon and delivers rain to the city's reservoirs may become considerably less reliable. Antonio Donato Nobre, a climatologist with Brazil's National Institute for Space Research, goes so far as to warn that if just 40 percent of the Amazon is cleared of trees there could be a massive and sudden shift of remaining forest to grassland and a breakdown of the current climate system. If deforestation continues, he said to the *New York Times* in October 2015, São Paulo will most likely "dry up."[6]

Leaving aside global climate change, the clearing and degradation of the Amazon rainforest is among the biggest disruptions of the natural water cycle that has occurred so far. When coupled with climate change, it portends profound consequences for Brazilian society that will ripple out to the world. And while Amazonian deforestation is unique in its scale and impact, the degradation of watersheds around the world poses numerous threats to communities—including heightened risks of fire, floods, soil erosion, pollution, and water shortages.

Watersheds function as nature's water factories. When operating well, they collect, store, circulate, and treat water as it flows through and across the landscape to join rivers downstream. But when degraded by poor forest management, unsustainable crop production, and urban development, they no longer deliver these services. Today, pioneering initiatives in China, Latin America, the United States, and elsewhere are demonstrating ways of repairing watersheds that benefit the communities and economies downstream. In so doing, they are cost-effectively securing the quality of drinking water, reducing risks of fire and floods, and improving the livelihoods of those living in degraded landscapes. They are showing that investing in healthier watersheds not only makes sense economically, it also builds resilience to the growing impacts of climate disruption.

~

I left Beijing, China's capital, on midnight train number 251, headed south for the city of Zhengzhou. Not long after we passed the illuminated Great Hall of the People, the rumble of the train lulled me into a deep slumber. I was determined to awaken early the next morning, so as not to miss the crossing of the Yellow River. The Chinese affectionately call the Huang He their mother river, because thousands of years ago it gave birth to the nation's earliest, thriving civilization. I traveled to China in June 1988 to learn more about the landscape that had given the Yellow River its name—the erosion of the yellowish soils of the vast Loess Plateau, an area one-and-a-half times the size of California in the river's middle and upper reaches. Little did I know I would soon see land experiments that would presage one of the biggest watershed restoration efforts ever undertaken on the planet.[7]

I awoke with more than an hour to spare. The countryside at dawn was already buzzing with activity—farmers in the fields harvesting their wheat, horse-drawn carts pulling loads of grain to market, and bicyclists heading to work down tree-lined dirt roads. Soon I caught a glimpse of the river's channel up ahead. As the train rumbled across it, I gazed below and was stunned at how little water it was carrying. At this point in its 5,464-kilometer (3,507-mile) journey, large diversions for agriculture and growing cities had turned the great Yellow River into a paltry trickle. Farther downstream, I later learned, the river was drying up about 130 kilometers (80 miles) before reaching the sea. By the end of the dry season, China's mother river was quite literally tapped out.

But the Yellow had another nickname that told a story not of shortage but of floods: China's Sorrow. Historically, the river transported 1.6 billion tons of silt each year, more than 90 percent of it coming from the highly erosive Loess Plateau. The river dropped much of that sediment load when it reached the flatter North China Plain, which resulted in the river gradually becoming elevated above the surrounding land. This posed enormous flood risks. Over a period of 2,000 years, the Yellow breached its dikes more than 1,500 times, causing devastating floods and loss of life.[8] As the populations of Beijing and other cities on the north plain expanded, concern also grew about the massive sandstorms that periodically choked the capital's air, as well as the buildup of sediment in downstream reservoirs.

After a brief visit with the Yellow River Conservancy Commission in Zhengzhou, I headed west to Shaanxi Province, in the heart of the Loess Plateau.

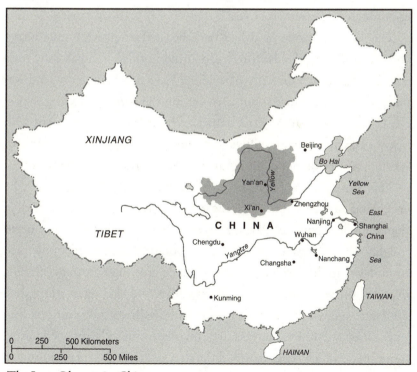

The Loess Plateau in China.

My traveling companions, Cheng Guangwei with the Loess Plateau Integrated Survey Team and my highly knowledgeable interpreter Zhang Junzuo, both worked at the Chinese Academy of Sciences. Shaanxi is perhaps most famous as home to the Terracotta Warriors in the ancient capital of Xian, but less famously it is also home to the most serious soil erosion problems in all of China. About half of the Yellow River's sediment load, some 830 million tons a year, originates from the lands of Shaanxi. Traveling around the region, it's easy to see why: there's almost no flat land and far too little vegetation on the steep slopes to hold the soil in place. Between 50 and 80 percent of the province's rainfall comes during three months, July through September, and the intense monsoonal storms wash the sediment off the hillsides. In northern Shaanxi, wind and water were eroding an average of 10,000 tons of soil from each square kilometer every year.[9]

This brown, barren land had once boasted extensive forests mixed with beautiful grasslands. It had supported not only China's early societies but also a rich diversity of wildlife. Scientists believe the land was fairly stable for centuries, but then during the Ming and Qing dynasties (1368–1644, 1644–1912, respectively) it spiraled downward with heavy tree cutting, crop cultivation, and overgrazing. In the twentieth century, a growing population put even more pressure on the land.[10]

This history was in plain view in a Shaanxi county, Mizhi, which covers an area of 1,212 square kilometers (468 square miles), a bit smaller than sprawling Phoenix, Arizona. Its people, who at that time numbered roughly 170,000, were some of the poorest in China. They worked from dawn to dusk eking out an existence growing crops and herding livestock on lands dissected by deep gullies and sloping at more than 20-degree angles. Most lived in tidy cave dwellings built into the hillsides. In an attempt to increase their income, some villagers tried raising more goats, which only worsened the devegetation. With fewer roots to hold the soil and capture the rain, erosion and desiccation increased—a reinforcing cycle of poverty and land degradation that seemed impossible to break.

The afternoon we arrived in Mizhi, after a five-hour drive on bumpy dirt roads from Yan'an, where Mao Zedong and Zhou Enlai spent several months planning the 1949 revolution, we got out of our car briefly to stretch our legs. A herder approached with his goats and then turned up a dirt road toward some cave dwellings. Zhang, my translator, asked if I'd like to go up and talk with some of the villagers. We came upon a woman chopping potatoes outside next to a wood-fueled stove. Her family, in tattered and torn clothing, gathered around. They invited us into their modest dwelling and asked us to stay for dinner, an astonishing act of generosity given how scarce their food supplies undoubtedly were. On the walls hung large posters of Marx, Lenin, Stalin, and Mao, stalwarts of the collectivization movement that was already yielding to a more market-oriented economy in China's countryside. After a brief conversation, we thanked them for the dinner invitation, explained that we needed to travel on, and departed feeling enriched by our brief encounter with this materially impoverished family.

While in Mizhi, we visited two villages. The first, Daichua, had been designated an experimental restoration village in 1979, soon after the Chinese government had instituted the "responsibility system," which gave farmers new incentives to improve agricultural productivity. Daichua resembled a fairy-tale village. It was eye-popping green, with trees, shrubs, and crops across the landscape. Where a deeply incised gully would have been, there was flat farmland 5 kilometers (3 miles) long. The villagers had built an earthen dam at the end of the gully to trap the eroding sediment. Over time, it built up a thick horizontal layer of soil, which the villagers called dam land. Nearby sat a rectangular pond to collect and store irrigation water, which greatly boosted yields on this newly created cropland.

The villagers had also shaped the sloping hillsides into flat terraces,

each about 5 meters (16 feet) wide. Hundreds of terraces planted in mil-
let, corn, sorghum, and soybeans lined the hillsides. On the unterraced
slopes, they had planted shrubs useful for fuel and animal fodder and
that held the soil in place. Fast-growing poplar, locust, and elm trees
waved across the landscape. Zhang, Cheng, and I hiked to the top of a
hill to get a better view. If more of the Loess Plateau could receive this
kind of treatment, I thought, the land and people's livelihoods might be
transformed.

That transformation was in fact beginning with the experimental work under way in Mizhi. With support from the United Nations
World Food Program (WFP), scientists at the Mizhi Experiment Station worked with villagers from 1979 to 1985 to terrace the land and
to substitute trees and grass for crops on land sloping more than 20
degrees. The goal was to reduce both the county's ecological degradation
and its poverty level. In return for their hard labor, villagers were typically compensated in grain from the WFP. Investments averaged about
210 yuan per hectare ($56 per hectare at the 1988 conversion rate).
Altogether, the WFP project area encompassed 750 square kilometers
(290 square miles), about 60 percent of the county. It included 241 villages, 21,340 families, and a total population of 105,400.[11]

Quanjiagou, the second experimental village we visited in Mizhi, had
the same Garden-of-Eden feel as Daichua. The conversion of sloping
cropland to trees and grass had dramatically changed the land-use pattern: cropland area was down by half, while the area covered in trees
was up by 85 percent and that in grass by more than 400 percent. But
even with only half as much cropland, the villagers harvested 17 percent
more food. With the extra revenue from animal husbandry and high-
value fruits and other tree products, per capita income during the six
years of the program had more than doubled.[12]

Granted, this was the China of three decades ago, and though my
trip was not scripted, it was carefully planned by the Chinese Acad-

emy of Sciences. In all likelihood there were failed efforts I did not see. But these pilot restoration efforts seemed to be a striking success. They had shown that the rehabilitation of one of the most erosive, incised, and barren landscapes on Earth was possible, and that it could greatly improve people's livelihoods and incomes at the same time. Of course the big question was whether the impressive outcomes I saw in Daichua and Quanjiagou could be sustained and replicated. Could this rehabilitation be scaled across enough of the Loess Plateau to benefit not only the local people but also those living downstream on the plains of the Yellow River?

Five years after my visit, an unexpected event helped spur that very attempt to scale up. In 1993 a sandstorm, dubbed the 5 May Black Wind, tore through a portion of the Loess Plateau and Inner Mongolia, killing scores of people and tens of thousands of farm animals, and causing massive damage to trees, crops, and property. The very next year the Chinese government began spending heavily to rehabilitate a wider area of the Loess Plateau. Between 1994 and 2005, with financial assistance from the World Bank, the government invested approximately $500 million in about 1,100 small watersheds across the region.[13]

The program became one of the early models of "payments for ecosystem services," an idea that would soon gain popularity in much of the world. It subsidized the cost of seeds and seedlings to the tune of $122 per hectare and paid villagers $49 per hectare per year for the erosion control and other ecosystem services the land rehabilitation would deliver. These subsidies were crucial to motivate farmers to convert their cropland to trees and grass. In 1999, midway through the decade of restoration, the Chinese government aligned the Loess Plateau restoration with its innovative Grain to Green Program, a vast land restoration effort through which the government pays millions of rural families to convert unstable croplands to trees and grass.[14]

As in the villages in Mizhi, the technical components of the program focused on terracing, tree planting (including higher-value orchards), grasslands, sediment-control dams, and irrigation ponds. But an important new incentive was added: the granting of land-use contracts. Every field or section of a terrace was contracted to a family, which then had rights to what that field produced, as well as the responsibility to maintain it. Around 1999, the government banned tree cutting and crop cultivation on sloping land. It also banned unrestricted livestock grazing. Villagers had to confine their sheep, goats, and cattle so the animals no longer destabilized the hillside soils. Not only did the ban prove crucial to the land's rehabilitation, it unleashed new livestock enterprises—including the sale of Kashmir wool and dairy products—that diversified the local economy and lifted incomes.[15]

By most measures the program proved a resounding success. According to researchers Kathleen Buckingham and Craig Hanson with the World Resources Institute, a nonprofit research organization based in Washington, DC, by 2006, per capita grain production had increased by 62 percent, even as the cultivation of erosion-prone, sloping land declined by 38 percent. Communities in one watershed saw their incomes more than double. Downstream benefits materialized as well. The volume of sediment entering the Yellow River dropped by approximately 300 million tons per year, reducing the cost of flood protection in the lower basin.[16]

Filmmaker John Liu, who captured the Loess Plateau's restoration in a compelling video, found that as the land was restored, wildlife and biodiversity returned. Villagers previously resigned to a life of drudgery and insecurity now had a spirit of optimism toward the future. "Before my dream was just to live in a brick house," said one of the villagers who spoke to Liu. "Now it's as if our dreams couldn't keep up with the changes."[17]

All told, the China–World Bank project alone restored close to 1 million hectares (nearly 2.5 million acres), but the benefits reached much farther. China has deployed the strategy developed by its scientists and the World Bank throughout the Loess Plateau region. As of 2008, more than 24 million hectares (59 million acres)—half of the plateau's degraded area—had been restored.[18]

The sustainability of this success, however, is not guaranteed. Only one-fourth of the 400,000 pine trees planted in northern Shaanxi Province survived. Those still living suck up so much moisture that water flows into the Yellow River could diminish. Chinese scientists are now reevaluating the full impacts of the tree-planting aspects of the restoration. They are also rethinking the use of check dams, the structures built across the gullies to trap sediment and control erosion, since in some cases the dams could increase flood risks. But incorporating the lessons of so-called failures is what adaptive management is all about. The design of future rehabilitation projects can incorporate the lessons learned from past efforts. More difficult, though, may be sustaining the payments villagers need to carry out the work of land stewardship. The government's payments have been set to expire in 2018.[19]

Still, the transformation of the Loess Plateau appears unmatched in scale and ambition, and it shows what is possible with effective partnerships and financial commitments. Twenty years ago, the plateau was a classic example of the "tragedy of the commons," the idea put forward by ecologist Garrett Hardin that individuals sharing common ground but acting in their own interest bring ruin to all.[20] Today, the Loess Plateau stands as a testament not only to the possibilities of land renewal but also to the potential of incentive-based collective action. A water cycle broken by decades of poor land use is functioning again. The benefits to local communities—food security, higher and more diverse income streams, and a healthier environment in which to live—are joined by downstream benefits to Beijing and other cities in the form of reduced sandstorms, better air and water quality, and reduced flood risks. The challenge now is to lock in those benefits, and expand upon them.

~

While I was in remote parts of China with no access to news, I'd missed a cultural shift back home: *global warming* had become a household phrase. NASA scientist James E. Hansen had gone before a committee of the US Senate on June 23, 1988, and testified that the earth was warming due to human emissions of carbon dioxide and other greenhouse gases. The headlines that followed brought climate change into the public lexicon, but action to slow its progress would prove painfully slow in coming.[21]

Global warming is changing the planet in a myriad of unanticipated ways, and one of the more surprising is the rise of megafires. Severe droughts and higher temperatures are conspiring with accumulated brush and thick understories to produce uncontrollable forest fires. On average, fires in the United States now consume twice as much area per year as three decades ago, according to the US Forest Service. The year 2015 set a record, with 10 million acres (4 million hectares) consumed by fire, including vast areas in Alaska and the Pacific Northwest. Federal firefighting costs climbed to $2 billion, more than eight times the expenditures in 1985. The Canadian provinces of Alberta and Manitoba recorded more than twice as many fires in 2015 compared with their averages for the previous quarter century. In Alberta, where a catastrophic fire in early May 2016 caused the evacuation of some 88,000 people from the oil-sands town of Fort McMurray, experts say the fire season now begins a month earlier.[22]

Wildfires are an integral part of a healthy forest, providing nutrients to soils and promoting regeneration. But during much of the twentieth century, forest management prioritized the protection of timber resources and communities near the forests over the health of the forest itself. As a result, forests that might naturally burn once in every 5, 10, or 20 years were not permitted to burn at all. Without these periodic natural burns to clean out the dense vegetation in the understory, fires that did erupt began to blaze with unprecedented speed and intensity, especially in times of drought. For the suppliers of drinking water downstream, this shift in forest management has spelled trouble.

In late June 2011, great balls of fire leapt from treetop to treetop in the Jemez Mountains of northern New Mexico. Drought, heat, wind, and fuel buildup in the forest created ideal conditions for a raging wildfire. Sparked by an aspen tree falling onto a power line, the Las Conchas fire burned extraordinarily hot and fast: during the first 14 hours, it consumed 43,000 acres (17,400 hectares)—nearly 1 acre a second. Over the next five weeks, the fire continued to grow. The town of Los Alamos, home to 12,000 people and a federal nuclear research facility, was evacuated for a week. By the time firefighters contained the blaze in early August, it had burned through more than 156,000 acres (63,130 hectares), the biggest blaze by far in New Mexico's history.[23]

But some of the worst problems came after the fire. Summer thunderstorms dropped intense rains onto the burned forest. With little vegetation to hold the soil and thousands of dead trees dotting the landscape, the monsoonal rains sent massive flows of ash and debris coursing through the mountain canyons.

In late August, about a week after one of the summer's worst storms, I headed to Cochiti Pueblo, a village of 800 tribal members that sits just east of the Jemez Mountains at the base of Cochiti Canyon. There I met up with Phoebe Suina, a Dartmouth-educated environmental engineer then working with the private firm Adelante Consulting. For weeks she had been monitoring the canyon's flooding and overseeing the construction of barriers and other measures to safeguard the pueblo throughout the postfire monsoon season. The previous week, Suina told me, had been extraordinary. After a 2-inch (50-millimeter) rain, a mud-

An expanse of mixed-conifer forest in the Cochiti Canyon watershed in the Jemez Mountains of northern New Mexico lies scorched following the Las Conchas fire. Photo by Craig D. Allen/USGS.

flow as wide as a football field roared out of Cochiti Canyon, pulling down power lines, ferrying a 150-foot (45-meter) tree down the canyon, and producing waves of black water 15–20 feet (4–6 meters) high. The changes have been jaw dropping, she said. "You wouldn't know you're in the same canyon."

Much of the ash and debris flowing out of Cochiti Canyon got trapped in Cochiti Lake, a federal reservoir on the Rio Grande, which forced authorities to close it to recreational use. Meanwhile, floodwaters from other canyons flowed directly into the Rio Grande. Several times after big thunderstorms, scientists with the University of New Mexico measured extremely low oxygen levels in the river. As bacteria broke down the heavy loads of organic matter in the water, they consumed the river's oxygen, effectively creating dead zones where fish and other aquatic organisms could not survive.[24]

As the river carried those blackened, debris-laden waters toward Albuquerque, the water authority for New Mexico's biggest city had some tough decisions to make. First and foremost was maintaining the quality and safety of the city's drinking water. While the utility's chief operating officer, John Stomp, expressed confidence that the treatment system could handle the ash-laden water, he told the *Albuquerque Journal* that he was concerned about the high cost of treating the blackened water and the damage the ash might cause to the plant's equipment. As a result, the utility stopped taking Rio Grande water for a portion of July as well as much of September and October. Even into the fall, the river carried high loads of ash, which was now more concentrated due to the river's lower flows. To make up the supply gap, the city pumped more groundwater. Santa Fe, the state's capital, also stopped drawing from the Rio Grande that summer, choosing instead to deplete reservoirs filled from a different watershed.[25]

Economists at the University of New Mexico later estimated the total costs of the Las Conchas fire—including direct firefighting costs, as well as fire-related flooding, diminished water quality, and other damages— to range between $156 million and $336 million. The midpoint of that cost range, $246 million, translates to $2,150 per acre burned.[26]

This fire–flood–water quality conundrum is a growing concern throughout the United States. Firefighting now accounts for more than 50 percent of the Forest Service's budget, up from 16 percent in 1995. The rising cost of battling fires has drastically reduced the funds available for the agency's other work—including the rehabilitation of forested watersheds so as to reduce fire risks.[27]

Fire's threat to drinking water supplies is particularly worrisome in the western United States. About two-thirds of the West's water supply comes from forested land. A study by the nonprofit American Forest

Foundation finds that across the West some 33.8 million acres (13.7 million hectares), an area roughly the size of Iowa, are both at high risk of forest fires and located in watersheds that supply drinking water to cities or towns. About 60 percent of those high-risk lands are national forests or other public lands, while 40 percent are privately owned.[28]

After two big fires in six years, including the record-breaking Hayman Fire in 2002, Denver, Colorado, saw heavy rains and floods dump 1 million cubic yards (85,000 cubic meters) of sediment into its Strontia Springs Reservoir—four times the volume that had accumulated in the reservoir over the previous two decades. When Denver Water, which serves some 1.3 million people in Denver and the surrounding suburbs, estimated the extra cost of water treatment, the removal of sediment and debris, and infrastructure repair to exceed $26 million, the agency decided to get proactive.[29]

Rather than risk repeating those expenditures fire after fire, Denver Water entered into a partnership with the US Forest Service, which manages the forestlands that supply the city's drinking water and itself had incurred some $37 million in postfire expenditures. Called the Forest to Faucets Partnership, Denver Water and the Forest Service are investing $33 million, split evenly between them, in tree thinning, the creation of fuel breaks, and other forest rehabilitation measures in areas critical to Denver's water supply. Denver funds its share of the program with an increase in household water bills of 4 cents per thousand gallons used, which translates to roughly a 1 percent increase in the average household water bill. Without the watershed investments, however, Denver ratepayers would almost certainly see bigger increases in their bills as fires took a greater toll on the city's water system.[30]

Back in New Mexico, concern about fire's threats to drinking water was growing. The Las Conchas was the second megafire in just over a decade. In May 2000, the Cerro Grande fire started as a "prescribed burn," a fire intentionally set under controlled conditions to reduce the buildup of understory growth and flammable fuels on the forest floor. But the burn got out of control and ended up consuming some 19,000 hectares (47,000 acres) of forest in the Rio Grande watershed outside of Los Alamos. It held the title for the largest wildfire in New Mexico's history until the Las Conchas fire 11 years later. These two massive fires in northern New Mexico within such a short period of time, combined with the growing vulnerability of the state's water systems to these drought-fire-flood events, caught the attention of Laura McCarthy, director of conservation programs with the New Mexico office of The Nature Conservancy.

McCarthy brought just the right blend of experience and skill to the challenge of designing a solution to mitigate these threats. She had worked for the US Forest Service for a dozen years, and she credits her time assessing private timberlands in New England for her belief in the importance of looking at the whole forest, not just one piece. That work also taught her about the power of collaboration.

"That was really pivotal for me," McCarthy said over a cup of tea in Albuquerque in early February 2016. "I was like 28 years old. After that I got totally hooked on the collaborative process."[31]

A decade later, working for a nonprofit forest conservation organization in Santa Fe, New Mexico, McCarthy got to see the lessons of the Cerro Grande fire firsthand. She helped shape a response that included thinning trees and rehabilitating the Santa Fe watershed through a "payment for ecosystem services" model that involved the US Forest Service, the city of Santa Fe, the Santa Fe Watershed Association, and The Nature Conservancy. But it was not fully collaborative, McCarthy said, and it was too small-scale. "For the Rio Grande, you can't be that narrow and succeed."

By the time the Las Conchas fire broke out in 2011, McCarthy had been working for The Nature Conservancy for six years, initially in its fire program. The record-breaking blaze and its impacts on Albuquerque's water supply gave McCarthy the opportunity she needed to attempt a larger restoration effort. After two years of scoping the challenge and meeting with representatives of most of the key agencies, businesses, and organizations with a stake in the outcome, she called a meeting in April 2013. "Imagine my anxiety," she said. "Will anyone actually show up?"

The room, as it turned out, was packed. Representatives from the urban water utilities, the business community, the county, the state forestry agency, the US Forest Service, the state engineer's office, and others came ready to roll up their sleeves. The upstream–downstream connection made sense to people, McCarthy said. "The idea resonates."

With that meeting, the Rio Grande Water Fund was launched. True to the idea that the beneficiaries of healthy watersheds should pay for the ecosystem services those watersheds provide, the water fund seeks to raise money to rehabilitate forestlands where significant fire risks threaten downstream water supplies. The fund's goal is ambitious: to restore 600,000 acres (242,800 hectares) of forested watershed over the next 20 years.

At an estimated cost of $21 million per year—or an average of $700 per acre ($1,730 per hectare) over the two decades—securing the necessary funding is no easy task. But the fund's value proposition is that these payments will pay off in the long run, saving money from future damage. To examine that claim, McCarthy's team at The Nature Conservancy looked at the costs of a hypothetical 180,000-acre (72,845-hectare) fire burning half of the watershed above two critical water supply reservoirs. Forest thinning and other rehabilitation measures were estimated to cost $73 million to $174 million, while damage estimates—including the costs of fighting the fire, dredging the reservoirs to remove fire-related sediment and debris, extra water treatment, and damage to property—ranged from $104 million to $1.3 billion. So even if just one large wildfire burned in this critical portion of the Rio Grande watershed, the upfront investment in a healthier, more fire-resistant forest would save money.[32]

To reap these benefits, however, the rehabilitation must be done to scale, in the right locations, and deliver the intended results. Based on tools developed by the US Geological Survey, the Water Fund worked with scientists to develop a "rapid assessment method" to identify areas where flows laden with debris were most likely to occur. One concern of McCarthy's, however, is that, while models of fire behavior are quite advanced, the tools available to model watershed functions are not yet very accurate. When she asked a University of New Mexico water scientist to review a dozen different watershed models and advise her team on which to use, the answer, McCarthy said, "was kind of 'none of the above.'"

Undeterred, McCarthy and the Water Fund team decided to support research to improve our understanding of watershed functions with respect to fire. These studies are a work in progress, but will advance the science over the long run. The two critical questions, McCarthy said, are these: "Are we changing fire behavior? Are we influencing the trajectory of watershed response to fire?"

Even as scientists learn more, the Rio Grande Water Fund has left the gate. By late 2016, the fund had restored 22,000 acres (8,900 hectares) of forested watershed land. Local businesses are turning what had been overcrowded, spindly trees—ideal fuel for a fire—into useable lumber for construction and wood to heat homes. Walatowa Timber Industries, which sells latillas, vigas, firewood, and other wood products, plans to double its staff in the coming year, as the volume of wood it is processing has grown sixfold. The fund has more than 30 investors, including regional water utilities, government agencies, and corporations, as well as a diverse array of local supporters and contributors, from banks and breweries to private individuals and family foundations. Diane Ogawa, executive director of the PNM Resources Foundation, attributes the business community's interest in investing in the water fund to its "triple bottom line—clean water, jobs, and fewer economic disruptions from wildfire."[33]

When asked about the biggest roadblock to making the Rio Grande Water Fund a success, McCarthy said simply, "It's complicated." Every agency is on a different planning and funding cycle and has different constituencies. Progress is slow. Patience, it seems, must accompany perseverance. As the saying goes, if you want to go fast, go alone, but if you want to go far, go together.

Collaboration—going together—is the heart of the Rio Grande Water Fund's governance structure and theory of change. Virtually all of the people who came to that initial April 2013 meeting, and the agencies and organizations they represent, remain active partners in the Water Fund. More than 50 agencies, businesses, and organizations have signed a nonbinding charter that commits them to work together to solve the challenges posed by fire and watershed management.

"That's why people keep showing up," McCarthy said. "They have ownership of it. It's theirs."

~

Fire protection is only one reason that cities, businesses, and public agencies are working with scientists and conservation organizations to repair their watersheds. Another is to reduce the cost of treating drinking water. A healthy watershed with stable soils, ample vegetative cover, and good buffers to capture runoff from farmlands and cityscapes can filter out many pollutants more cost-effectively than a treatment plant can. A recent survey of 37 US treatment plants by the nonprofit American Water Works Association (AWWA) found that the conversion of 10 percent of a watershed from forest to housing or other developed uses increases water treatment costs by an average of 8.7 percent. An earlier analysis of 27 US water suppliers by AWWA and the Trust for Public

Land, a conservation organization based in San Francisco, California, found that when forest cover in a watershed dropped from 60 percent to 30 percent, average water treatment costs doubled; when it dropped from 60 percent to 10 percent, average treatment costs tripled.[34]

New York City, with a population of 8.4 million, is one of the biggest cities in the world to invest in nature to keep its award-winning drinking water clean. The city gets 90 percent of its water from the lands of the Catskills–Delaware watershed. US drinking water rules require any city that relies on rivers, streams, or other surface waters for its supply to build a filtration plant unless it can demonstrate that good watershed protection will be sufficient to meet the drinking-water standards. Since 1997, New York City has opted for nature's way of treating water over the technological solution—and has saved billions of dollars in the process.

A hallmark of New York City's watershed protection program is a memorandum of agreement (MOA), signed after many years of negotiation, by city, state, and federal officials; conservation organizations; and some 70 watershed towns and villages. The MOA committed New York City to invest on the order of $1.5 billion over a decade to restore and protect the watershed, as well as to measures that would improve the local economies and quality of life of residents living in the watershed. It is a complex partnership that's been tough to execute. Because about three-quarters of the Catskills–Delaware watershed is forested, but three-quarters of it is also privately owned, acquiring land has been a crucial component of the program. Legally the city could take high-priority watershed lands by eminent domain, but it committed in the MOA not to do this. Instead, it acquires land from willing sellers and pays full market price for it. And to prevent its land acquisitions from decimating the revenue base of watershed towns, it also pays property taxes on the land it buys.[35]

Between 1997 and 2013, the city had signed contracts with nearly

1,500 landowners to protect some 130,000 acres (52,600 hectares) either by direct acquisition or through conservation easements, a deal in which the city pays landowners to accept restrictions on how they may use their land. Altogether, including state lands and other protected areas, some 37.5 percent of the Catskill–Delaware watershed is protected from development. New York City has also upgraded some 4,600 septic systems, built three wastewater treatment plants, and undertaken forest and stream projects to control erosion and runoff. It has spent $1.7 billion on these various watershed protection measures, but it has managed to avoid the need to construct a filtration plant estimated to cost at least $10 billion. Even adding in the $1.4 billion the city spent to construct an ultraviolet disinfection plant to kill any pathogens that sneak through the watershed, it's an impressive return on investment.[36]

But it hasn't been easy. Tensions between the rural watershed towns and the big city have at times run high. The city is trying to do more to support upstate economic development, for example, by creating opportunities for recreation and tourism. As in China's Loess Plateau, even when a government entity foots the bill, there is no guarantee of success over the long term. Both the beneficiaries and the providers of watershed services need to feel that the deal not only delivers something of value to them but is also fair. So far, New York City's agreement with the people of the Catskills has held, and its watershed program has passed muster with federal drinking water authorities.

"We've come a long way since 1997 when the watershed agreement was signed," said Ira Stern, natural resources division chief of New York City's Department of Environmental Protection. "We've evolved over the last 20 years."[37]

Over the last two decades, Latin America has emerged as a world leader in the development of watershed funds to protect those lands that deliver reliable, clean water to cities and businesses downstream. Much of the credit goes to the pioneering actions of Quito, the capital city of

Ecuador and today home to some 2.6 million people. Quito gets about 80 percent of its drinking water from protected biological reserves that encompass 520,000 hectares (1,285,000 acres) of high-elevation grasslands and cloud forests. Although they are formally part of Ecuador's national park system, these reserve lands are also used for cattle, dairy, and timber production. On the order of 30,000 people live within or around the reserves.

With the biodiversity of the reserve lands at stake along with Quito's water supply, The Nature Conservancy helped establish a first-of-its-kind trust fund to finance watershed protection. In contrast to New York City's program, which is funded by the city itself, Quito's Fondo del Agua launched in 2000 with the idea of getting local businesses and agencies that stood to benefit from cleaner or more reliable water supplies to contribute. Once the fund was sufficiently capitalized, its managers would then disburse the interest it earned to watershed projects. Along with The Nature Conservancy and Quito's water company, initial investors included a brewery, a water-bottling company, and the hydropower company that supplies the city with electricity. The water fund has grown to $12–$14 million and pays out tens of thousands of dollars a year for projects such as fencing off livestock from streams, planting trees, educating local communities, and hiring guards to prevent illegal logging in the preserve lands.[38]

The success of Quito's Fondo del Agua has inspired numerous similar ventures throughout Latin America, including in Bogotá, Colombia, and Lima, Peru. The Nature Conservancy aims to help launch 32 water funds in the region, protecting some 3.6 million hectares (9 million acres) of land that filter and supply drinking water for some 50 million people.[39]

Researchers are designing better planning and assessment tools to help watershed programs achieve their potential. For example, a series of software models called InVest developed by the Natural Capital Project, a joint endeavor of Stanford University, the University of Minnesota, The Nature Conservancy, and the World Wildlife Fund, helps cities and regions incorporate ecosystem services into their land-use and development decisions. But as the inspiring efforts in the Loess Plateau, the Rio Grande valley, New York State, and Latin America show, there is no one recipe for building and sustaining a living, working watershed. It is often as much art as science and finance.

One lesson, however, is clear: collaboration and sustained commitment are essential. These may be hard earned, but the payoffs can be big—not just in dollars and results, but also in connecting people to the source of their drinking water and the landowners who protect it.

CHAPTER 4

Make Room for Floods

Every shopping center, every drainage improvement, every
square foot of new pavement in nearly half the United States
was accelerating runoff toward Louisiana.
—*John McPhee*

Floods occupy a distinct place in human mythology. In the lore
of several early religions, the coming end of the world is attributed to
a massive deluge. The story of Noah and the Ark as told in the book
of Genesis closely parallels another in the early Babylonian epic of Gil-
gamesh, in which a god forewarns Utnapishtim of a coming flood and
advises him to build a ship to save himself, his family, and the seeds of
all living things. Even as many cultures celebrate water through sacred
rituals as the source of life, water engenders fear in its capacity to rise up
and destroy. Water gives, but it can also take away.

It is this wild, dangerous side of water against which human civilization has struggled over the millennia. Early societies sprang up along

the Nile, Tigris-Euphrates, and Yellow Rivers in order to grow crops in the floodplains and ply their host rivers by boat to trade goods. According to Chinese legend, efforts to control the waters of the Yellow River began some 4,000 years ago, when a massive flood roared through the basin and inundated the plains below. A legendary hero named Yu organized workers to dredge channels so as to confine the raging waters. Because he tamed the flood and distributed the reclaimed farmland to the people, Yu earned the mandate to become founding emperor of China's first dynasty. Today, a giant statue of Yu the Great looms above the river's floodplain near the Yellow River Conservancy Commission in the city of Zhengzhou. With one hand gripping a tool used to tame China's "mother river," the imposing Yu simultaneously evokes control, hard labor, victory, and prosperity.[1]

But the Yellow River would not obey for long. During the 2,000 years prior to 1950, the unruly river broke through dikes along its banks more than 1,500 times. Among the world's deadliest disasters, excluding famines and pandemics, was the spate of flash floods in central China in 1931, when the Yellow, Yangtze, and Huai Rivers all overflowed, killing up to 4 million people.[2]

Floods are necessary to the life of a healthy river. They distribute nutrients and sediment downstream, rejuvenate riparian (riverside) vegetation, recharge groundwater, and cue fish to migrate and spawn. But growing civilizations have a love/hate relationship with this natural phenomenon. Being near rivers means access to drinking water, shipping lanes, electrical power, and recreation. It also puts cities and towns in harm's way. We have adapted to floods largely by trying to control them with dams, dikes, and levees. But as events over the last quarter century have made abundantly clear, we would do well to leave more room for rivers to be rivers.

～

The summer of 2002 changed Europe's relationship with its rivers forever. In early August, heavy rains triggered waves of flooding down the Danube in Austria and the Vltava, Labe, and Elbe Rivers in the Czech Republic and Germany. Cities along the Elbe River, which is called the Labe River until it crosses from the Czech Republic into Germany, were particularly hard hit. Some communities saw flood levels expected only once in 500 years. In the city of Dresden in the German state of Saxony, the Elbe peaked at a height of 9.4 meters (30 feet), topping the previous record set in 1845. The main railway line between Dresden and Prague was closed for more than four months. Even though volunteers and the military fortified defenses as the flooding unfolded, the raging rivers broke through dikes and seeped through saturated soils beneath the riverbank barriers. Many thousands of people were evacuated and vast areas were inundated, including nearly 600 square kilometers (230 square miles) in Germany alone.[3]

Much of central Europe was already reeling from floods before that epic summer. The Rhine had flooded big in 1995 and the Oder in 1997. In the summer of 1999, the Danube surged through Belgrade, forcing evacuations in portions of the Serbian capital. From 1998 through 2002, European countries suffered 100 major floods that collectively took 700 lives, displaced half a million people, and cost €25 billion ($26 billion).[4]

And the punishing floods just kept coming. In 2005 high waters in the Danube basin took 31 lives in Romania and washed away 600 bridges. The next year, an "extremely rare coincidence" of large floods in a number of tributaries in the basin led to a 100-year flood along more than 1,000 kilometers (620 miles) of the main river channel. The high waters destroyed hundreds of buildings in Bulgaria and Romania. In 2009, the river immortalized by Richard Strauss's lilting "Blue Danube" waltz inundated 110,000 hectares (272,000 acres) of Romanian farmland. In 2013, it surged to heights in the German city of Passau not seen

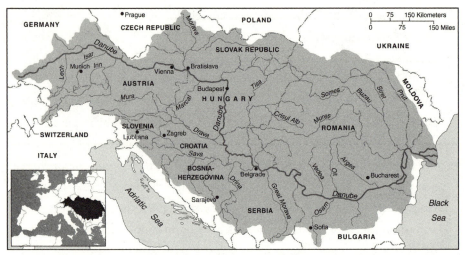

The Danube River basin in Europe.

since the Renaissance, and in Budapest it lapped at the back steps of the Hungarian capital's parliament building.[5]

The rivers of Europe were delivering a message: they want their floodplains back, and they will fight to get them. It's a message we in the industrial world are reluctant to hear, much less accept. But in Europe and other parts of the world, coalitions of scientists, conservationists, and government agencies are making the case for partially conceding this fight and giving rivers room to roam across their floodplains once again.

Throughout the nineteenth and twentieth centuries, the United States, Europe, and other rapidly industrializing regions invested heavily in dams and levees that intentionally disconnected rivers from their floodplains. (Levees and dikes are engineered embankments, usually constructed of earth materials, built along a river to prevent it from overtopping its banks.) The idea was to control floods so that engineers could manage rivers for shipping, energy production, and water supply, while protecting lives and property from high waters. Safeguarded by levees, cities and farms could move into former floodplain areas, where they would have access to the river and to the rich soils built up by the river's sediments over the centuries. In this way the Danube, Europe's second-largest river, and the lower Mississippi in the United States lost 80 percent of their original floodplains.

To planners and engineers, this massive network seemed to be a sensible way to put rivers to work. There was just one problem: rivers would not get with the program. When levees confine a river into an artificially narrow channel, floodwaters will rise higher and flow faster than they would if unrestrained. Seeking to spread out, they will press hard against their artificial embankments until they find a weak spot, and then burst through. As residents along the Elbe discovered in 2002, floodwaters can also seep beneath levees through saturated soils and then inundate land on the other side. Often rivers will take both escape routes simultaneously. The vast majority of big floods in central Europe over the last two decades have occurred along rivers specifically engineered to do away with them.

In response, Austria, Germany, the Netherlands, Romania, and other European countries began to put nature, or "green infrastructure," into their flood-control strategies. The idea was not only to reduce flood damages but to reclaim some of the benefits lost from straightjacketing their rivers. When a river spreads across its floodplain, it not only slows down; it nourishes riverside habitat and recharges groundwater. Some of that groundwater will slowly return to the river and sustain its flow during drier times. The vegetation, soils, and microbes that make up the floodplain ecosystem cleanse the river's water as it soaks through the earth. It is a relationship of reciprocity: the river delivers water to the floodplain, and the wetlands purify and store that water for the river's later use. Habitats get restored for fish and wildlife that depend on the nutrient-rich floodplain for nesting, breeding, feeding, and growth.

We rarely put a dollar value on the services floodplains provide, in large part because markets don't price them. It's a Catch-22: because the economic value of floodplains is hidden, they get drained and converted to farmland and suburban communities at a rapid rate. A research team led by Robert Costanza, a professor of public policy at Australian National University in Canberra, reports that between 1997 and 2011 the global area of floodplains and swamps declined by 63 percent, and that the value of the ecosystem services lost as a result totaled some $2.7 trillion per year, equal to 3.6 percent of the 2011 gross world product.[6]

Many people in the developing world depend on natural floodplain services for their very livelihoods. The Bodi, Karo, Kwegu, and Mursi tribes in eastern Africa, for example, rely on the natural flood cycles of the Omo River, which flows through Ethiopia into Kenya, for their sustainable practices of flood-recession farming, fishing, and livestock grazing. Like generations of their forebears, they plant sorghum, maize, and beans in the riverside soils after the yearly flood, making use of the moisture and nutrient-rich sediment deposited by the Omo's floods each year. Their way of life now hangs in the balance, however, as Ethiopia's giant Gibe III Dam replaces the river's natural rhythms with flows regulated to optimize the production of hydropower and the delivery of irrigation water to industrial farms. A more holistic approach might avoid the errors industrialized countries are now trying to correct and instead balance the benefits of hydropower and flood control with critical floodplain services.[7]

Ironically it was the Netherlands, with its highly engineered seawalls and popular folklore of the little Dutch boy who plugged a dike with his finger, that inspired a movement to rethink flood control in Europe. When levees failed during the big 1995 flood on the Rhine, large portions of the Netherlands were inundated. The low-lying nation, which is home to some of the world's top water management experts, made a decision not only to shore up its engineering defenses but to figure out where and how to give the river some space to flood.

That notion turned into a $3 billion program of some 35 projects to lower dikes, set others farther back from rivers, create spillways, and turn some farmland back to floodplain. Called ruimte voor de rivier (room for the river), the idea is to help rivers safely carry "a lot more water than they do now," according to a spokeswoman for the program. One project along the Waal River in the eastern part of the country, for example, is converting land once slated for development into a park that can receive and store floodwaters when necessary. Another project, south of Amsterdam, involves lowering a dike along a low-lying area the Dutch call a polder and allowing floodwaters to spill into it. Authorities compensated farm families that had to vacate the polder, judging the project's benefits—including lower risk of flooding for some 140,000 residents downstream—to be well worth the cost.[8]

In a vast watershed like the Danube, reconfiguring flood control is far more complicated. The fabled river's watershed spans portions of 19 nations, and on its 2,840-kilometer (1,765-mile) journey from Germany's Black Forest to the Black Sea it courses through 10 countries. The challenge was pushed forward, however, by the European Union's 2000 Water Framework Directive, which calls on all EU countries to assess the health of their rivers and then restore them to at least "good" condition. In 2007, after so many years of extraordinary flooding, the European Parliament and the EU's Council of Ministers set out a specific directive on floods that requires member countries to assess their risks, develop maps of those risks, and then draw up plans for mitigating flood damages. Those plans are to include measures to reconnect rivers to floodplains and to limit home and building construction in flood-prone areas. In the case of international river basins like the Danube, the EU called upon countries to "coordinate so that problems are not passed on from one area to another."[9]

Many EU countries are hard at work turning the flood directive into on-the-ground projects. In the Danube basin, the Czech Republic, Slovakia, and Austria are reestablishing meanders in the heavily straightened and channelized Morava River. Austria has reconnected several stranded meanders with the main channel of the Danube, and along with Germany is removing levees to reconnect the Inn River, the Danube's third-largest tributary, to its floodplain. Like the Netherlands, Germany and Romania are using low-lying polders to retain floodwaters.[10]

Such projects take time to design, gain public acceptance, and implement. By 2015, only 3 percent of the roughly 193,500 hectares (478,000 acres) of floodplain in the Danube watershed identified as potentially ripe for reconnection to their rivers had actually been rejoined; an additional 21 percent had been partially reconnected. Plans are in place to reactivate more than 15,000 hectares (37,000 acres) of floodplain by 2021.[11]

Some of the most ambitious efforts are taking place in the Danube Delta, Europe's largest wetland ecosystem and home to some 320 species of birds. The Green Corridor for the Danube, a program launched in 2000 and spearheaded by the World Wildlife Fund, estimates that the floodplains of the lower Danube annually deliver benefits worth some €500 per hectare ($224 per acre). Ukraine has bulldozed dikes on the delta islands of Tataru and Ermakov, restoring natural flooding to 800 hectares (1,980 acres) and vital habitat for white-tailed eagles, pygmy cormorants, and other rare birds. The inner lakes now serve as spawning sites for fish from the Danube. Together, the four lower-Danube nations—Bulgaria, Moldova, Romania, and Ukraine—have protected some 1.4 million hectares (3.5 million acres) of wetlands, providing flood control, water security, and recreation.[12]

~

As in Europe, a barrage of big floods has also forced the United States to rethink its approach to flood management. Between 1980 and 2013, properties in the United States racked up more than $260 billion in damages connected to flooding. While a good share of this total was due to seaside flooding from Hurricanes Katrina along the Gulf Coast and Sandy along the eastern seaboard, inland flooding claimed its portion as well. Indeed, large floods hit midwestern and Gulf states in 1993, 2008, 2011, 2013, 2014, 2015, and 2016. In mid-August 2016, Louisiana was socked with its second catastrophic flood in six months. Some 30,000 people were rescued from the floodwaters, which damaged 40,000 homes. The US National Oceanic and Atmospheric Administration (NOAA) classified the Louisiana disaster as the sixth "1-in-500-years event" within the previous 12 months. Put another way, those six events—which included floods in Maryland, South Carolina, Texas, and West Virginia, as well as the two in Louisiana—each had less than a 0.2 percent chance of occurring during that year.[13]

In the Mississippi River basin, which drains 41 percent of the contiguous United States, the flood of 2011 was one for the record books. More floodwater surged through the basin that spring than during the historic flood of 1927, the most destructive river flood in US history. During the 1927 event, the river breached or overtopped levees in 145 locations, inundated 16 million acres (6.5 million hectares) of cities and farms, and displaced 700,000 people from their homes, some for weeks or months. That flood drove home an important lesson to early twentieth-century engineers: a "levees-only" strategy for flood control does not work against a big flood—at least not in the Mississippi, the third-largest watershed in the world, after the Amazon and the Congo.

Since its creation in 1802, the US Army Corps of Engineers (the Corps) has been the nation's premier river developer. A young Congress first called on the Corps to expand shipping and commerce by modifying rivers and harbors. Before long the Corps's portfolio expanded to include the construction of dams for hydropower, water supply, and flood control, as well as levees to make floodplains safe for growing farms and cities.[14]

Following the disastrous experience of 1927, Maj. Gen. Edgar Jadwin, then chief of the Corps, urged the US Congress to adopt a new flood policy, which it did the very next year. The Flood Control Act of 1928 instructed the Corps to stop piecemeal management of floods in the Mississippi basin and to adopt a coordinated system. That system should be designed to accommodate the largest flood believed to have a reasonable chance of occurring, the so-called design flood. While levees would remain the system's backbone, the act called for something new: the creation of "floodways"—areas where some water from a raging Mississippi could exit the main channel and flow into a designated area on its former floodplain. In that way, some pressure would be taken off the levees, reducing the risks of a breach. Families and farms occupied parts of these floodways, however. While the government purchased easements from the landowners, the intention was to use these exit ramps for the river only during high-risk floods.[15]

The massive flood of 2011 put the Mississippi flood-control system to the test like no other event during the preceding eight decades. In early May 2011, rapid melting of record snows combined with rainfall levels six to ten times higher than normal to send a surge of high water down the valley. The Mississippi's volume and height both broke records at dozens of river gauges from Cairo, Illinois, where the Ohio River joins the Mississippi, to Baton Rouge, Louisiana, near the Gulf Coast. For the first time, three of the system's four floodways were called into action during a single flood.

For Maj. Gen. Michael Walsh, then commanding general of the Corps's Mississippi Valley Division, the floodway option forced a tough decision. The high waters surging through the upper Mississippi and its tributaries placed the properties and lives of the 2,750 residents of Cairo, Illinois, at risk. To save Cairo, Walsh directed his engineers to blast open the levees at the Birds Point floodway. The river's level quickly dropped a couple of feet, sparing Cairo. But opening the levees inundated some

130,000 acres (52,600 hectares) of farmland and forced the evacuation of 100 homes. At a news conference afterward, Walsh said that he personally knew many of the people who lived and worked in the flooded zone, and considered them friends. Blasting open the levees at Birds Point, he said with a military man's restraint, "was a difficult decision."[16]

But by almost any reckoning, the new system and the engineers who operated it passed the tough test of 2011 with flying colors. Astoundingly, no one died in the flood. While it damaged some 21,000 homes and buildings, the Corps projected that under the old levees-only system some 1.5 million structures would have been damaged or destroyed. Economic losses totaled $2.8 billion, but the $14 billion federal investment in floodways, levee upgrades, and other components of the system saved an estimated $234 billion in damages.[17]

Clearly, the floodways were successful during an extreme storm. But what about giving the river access to some of its floodplain permanently? In addition to preventing flood damage, floodways could purify and store water, create habitat for fish and wildlife, sequester carbon, and provide a place for fishing, hunting, and recreation. Using this approach, ecologists and hydrologists would join their knowledge of wetlands and floodplains with engineers' expertise about levees and infrastructure.

Scientists, conservationists, and government agencies are beginning to explore these options. Their efforts are part of an evolving eco-engineering approach that works *with* nature's ability to control floods rather than against it. As in Europe, the aim is to make river systems healthier while mitigating flood risks.

Determining where and how to start is not an easy task. In the lower Mississippi River, which extends 954 miles (1,535 kilometers) from its confluence with the Ohio River near Cairo, Illinois, to the Gulf of Mexico, the levee system has shrunk the original floodplain by 80 to 90 percent and cost the river system some 23 secondary channels spanning nearly 10,000 acres (4,000 hectares). In the two decades following the 1927 flood, the Corps also cut the river off from 16 of its meanders. Instead of taking a slow and sinuous path to the sea, the river now flows fast and straight. That change benefited ship navigation but was terrible for the natural ecosystem. Cutting off the meanders not only shortened the Mississippi by 152 miles (245 kilometers), it caused the faster-flowing river to erode its own bed and banks. Stranded from the river, the meanders could no longer sustain habitats vital to fish, birds, and wildlife.[18]

For conservationists and like-minded engineers, finding ways to reconnect the Mississippi to parts of its historical floodplain is a high priority. The exchange of water, nutrients, sediment, and living organisms between the two is critical to nutrient cycling and the ecosystem's productivity, as well as to restoring habitats crucial for the life cycles of many species. Young fish, for example, take refuge in floodplain wetlands during floods, and an ancient-looking fish called the alligator gar spawns in the lower Mississippi's riverside swamps. Snowy egrets and other wading birds nest and feed on the lower Mississippi's floodplain, while populations of black bear, including the threatened Louisiana black bear, rely on the active floodplain's thick forests.

In 2012 the Lower Mississippi River Conservation Committee (LMRCC), a nonprofit coalition of 12 agencies from Arkansas, Kentucky, Louisiana, Mississippi, Missouri, and Tennessee, joined with the nonprofit Mississippi River Trust and the US Department of Agriculture's Natural Resources Conservation Service (NRCS) to restore wetlands and forests within the lower river's active floodplain, or batture. The Lower Mississippi River Batture Reforestation Project identifies areas that are likely to flood, works with interested landowners to create easements on these lands, and then plants trees to reestablish bottomland forests. The program aims to expand habitat for fish and wildlife, diversify local economies through more recreational opportunities, and reduce US taxpayer–financed disaster relief. The project area spans 345,800 acres (140,000 hectares) of cropland, a good portion of which was under water in late summer 2015. At that time, 60 properties totaling 13,427 acres (5,434 hectares)—more than half of them in Tennessee— were either enrolled in the program or in the application process.[19]

Another project known as Island 70, spearheaded by the LMRCC, involved notching a dike north of Rosedale, Mississippi, to restore flow to 3.5 miles (5.6 kilometers) of a side-channel. The team expects two endangered species, the pallid sturgeon and the interior least tern, to benefit from the restored habitat. As of 2015, the LMRCC and its part-ners, which include the Corps, had completed 14 projects that together have restored 56 miles (90 kilometers) of channel habitat and thousands of acres of surrounding land.[20]

Meanwhile, The Nature Conservancy is working with the US Fish and Wildlife Service and other partners to reunite the Ouachita River with 16,000 acres (6,500 hectares) of former floodplain forest. A beau-tiful tributary in the lower Mississippi basin named by early French settlers, the Ouachita flows out of Arkansas into Louisiana. The restora-tion team is removing portions of a 17-mile-long, 30-foot-high levee to re-create floodplain habitat while reducing flood risks downstream.[21]

Restoring the upper Mississippi basin is also critical. Over the last three-quarters of a century, while engineers were busy building locks, dams, and levees, vast areas of water-absorbing wetlands were being drained and filled to make room for farms and homes. Today over half of the corn and 47 percent of the soybeans grown in the United States come from the upper Mississippi River basin. But this productivity came at considerable ecological expense: Illinois, Indiana, Iowa, Mis-souri, and Ohio have each lost more than 85 percent of their wetlands. Altogether, the states of the upper Mississippi basin have converted 35 million acres (14 million hectares) of wetlands, an area the size of Illi-nois, to farms and towns.[22]

Restoring even a portion of these wetlands, which absorbed and held water like a sponge, could protect against extreme floods. In all too many locations, however, trends are going in the opposite direction. The Great Midwest Flood of 1993, when river levels set records at 92 locations and floodwaters broke through or overtopped more than a thousand levees, should have been a wake-up call to limit development in key parts of the floodplain. Instead, substantially more flood-prone areas got developed. According to Nicholas Pinter, professor in applied geology at the University of California–Davis, within a dozen years, 28,000 new homes and 6,630 acres (2,680 hectares) of commercial and industrial development had been built on land that was under water in 1993. In the St. Louis area alone, new development worth $2.2 billion had been situated on land that was inundated in 1993.[23]

Sensible zoning must clearly be part of any comprehensive strategy that combines floodplain restoration with engineering infrastructure across the Mississippi watershed. Such a system would not only be better at preventing flood damage, it would be better for the environment and the economy. For most flood-control managers, however, this is a new way of thinking. To help the process along, LeRoy Poff, an ecologist at Colorado State University in Fort Collins, Colorado, and several colleagues developed a method called eco-engineering decision scaling (EEDS). It's a cumbersome name, but EEDS does something few if any water management tools yet do: it integrates ecology and engineering while designing for a future of climate change.

"Traditionally, ecologists have had the role of 'fixing' a degraded ecosystem after a project has already been built," Poff said upon publication of the team's work in 2015. "It's time that decision-makers and engineers try something different and invite ecologists to the table." The real novelty of the work, Poff told me when we met up in Fort Collins in March 2016, "was putting it in the realm of climate risk."[24]

When it comes to flood protection, factoring in climate change is critical. Because a warmer atmosphere can hold more moisture, climate scientists expect the future to bring more frequent and intense storms. So far it's been difficult to pin specific floods on global warming; too many other factors are at play. But in a new area of science called event attribution, researchers are analyzing extreme weather data to determine whether some events would have been so unlikely to occur without the influence of human-induced warming that a climate fingerprint can be assumed, or at least considered highly likely. For example, England and Wales experienced widespread flooding in October and November 2000, which turned out to be the wettest autumn months since record keeping began in 1766. After running several thousand model simulations with and without the influence of global warming, Pardeep Pall of the University of Oxford in the United Kingdom and his colleagues concluded that "it is very likely that global anthropogenic greenhouse gas emissions substantially increased the risk of flood occurrence in England and Wales in autumn 2000."[25]

Such cautious attribution doesn't turn climate change into a smoking gun. But a 2016 review of attribution studies by the US National Academies of Sciences, Engineering, and Medicine cites findings that "it is likely" that since 1950 the number of big storms over land has increased in more places than it has decreased, and "confidence is highest for North America and Europe." As for the near-term future, "the frequency and intensity of heavy precipitation events over land will likely increase on average."[26]

Risk is all about probabilities. Only a foolish planner would ignore such findings, however carefully worded. But there's more. As temperature rises, scientists "robustly" expect the air to hold more water at a rate of about 6–7 percent per degree Celsius, and "a simple hypothesis" is that the intensity of heavy rainfall will increase at that same rate. Some research suggests that what were 1-in-20-year events in the 1950s are becoming 1-in-15-year events.[27]

Which takes us back to Poff's decision tool. To demonstrate how it works, he and his colleagues use an example from the Iowa River. Their models show that higher levees around Iowa City, combined with an intentional flood release from the region's dam, would make the system more resilient to rapid climate change and sustain the river's health. Returning water to the floodplain would benefit plants and wildlife, while also freeing up room in the reservoir either to capture more flood-waters or to use a portion of the "flood pool" for other purposes. It's an optimization analysis familiar to engineers, but one that puts the environment on par with engineering, and that frames decisions in the context of climate risks.[28]

Napa County, California, took this kind of approach when redesigning its flood-control system for the Napa River. The river flows about 55 miles (88 kilometers) from Mount Saint Helena through California wine country and empties into San Pablo Bay. In the early twentieth century, engineers straightened and deepened the Napa's channel and filled in its wetlands and tidal marshes. After enduring 11 serious floods between 1962 and 1997, local officials and residents knew they needed to do something different. "The Army Corps of Engineers wanted to build a concrete ditch down the middle of Napa," said county supervisor Diane Dillon, "and no one here wanted that."[29]

The county asked the Corps instead to collaborate on a "living river" strategy—a series of projects that would protect Napa from the 100-year flood but also enhance the natural habitat of the river. In 1998, residents voted in favor of a half-cent increase in their local sales tax to pay their share of the $366 million effort to reconnect the Napa River with its historical floodplain, move homes and businesses out of harm's way, revitalize wetlands and marshlands, and construct levees and bypass channels in strategic locations. "Since the start of the project I think we've had a billion in new private investment in the city of Napa, so that's been a huge help economically," said the city of Napa's mayor,

Jill Techel. "The wetlands have been restored and so we now have flora and fauna that has come back and enriched us." New trails give residents a place to bird-watch and hike.[30]

Some creative detective work by the nonprofit San Francisco Estuary Institute has helped inform the restoration. Researchers gathered clues about the Napa River's history from sources as obscure as landscape paintings, diaries kept by pioneers, and maps from early Spanish explorers. A fur-trapper's journal from 1833 mentioned beaver sightings, so the restoration team decided to allow the hardworking rodents back to build their dams, which create pools that often serve as nurseries for young fish. The goal is not so much to re-create the old look of the river as to restore ecological functions and build more resilience in the face of a changing climate. By the time the full suite of projects is complete, more than half of the Napa River's main channel will be restored.[31]

Sixty miles northeast of Napa, just outside of downtown Sacramento, California, a sign appears for the "Yolo Bypass." It sounds like the name of a highway that routes you around a city, but in fact it's a floodway that connects the Sacramento River to its floodplain. Like the floodways on the lower Mississippi, the Yolo was created in the 1930s after it became clear that levees alone could not protect the growing state capital from flooding. But unlike the Mississippi floodways, the Yolo Bypass gets used quite regularly, maybe six years in every 10. Its deployment requires no weighty decision like Maj. Gen. Walsh had to make about blowing open the levees at Birds Point. When the Sacramento rises to a certain level, it simply spills over a weir into a 59,300-acre (24,000-hectare) floodplain area. In March 2016, as heavy rains in northern California sent floodwaters down the Sacramento Valley, farmers with land in the floodway were warned ahead of time to get their equipment and livestock to higher ground. The bypass not only helps control flooding, it helps replenish California's depleted aquifers.[32]

The Yolo Bypass is a critical piece of infrastructure—green rather than gray infrastructure—in the region's flood-control system. During large storms, it might convey 80 percent of the Sacramento River's floodwaters out of the channel and onto the floodplain. More than a quarter of the bypass is a designated wildlife area used by nearly 200 species of birds. The California Department of Fish and Wildlife, which manages the area, permits hunting of waterfowl, pheasants, and doves. Together with the Yolo Basin Foundation, it also runs a "Discover the Flyway" program, which 45,000 students have visited to learn about wetlands, wildlife, and agriculture in the Central Valley.[33]

With the risk of flooding rising in much of the world, the case is strong for cost-effective systems that not only protect against damage but also reap the benefits of healthy floodplains. This will require cadres of ecologists to join civil engineers in shoring up flood defenses. And giving rivers room to roam will take smart policy and strong leadership, not to mention money, at all levels of government. But an eco-engineering approach is an important advance in our civilization's long relationship with floods. China's legendary Yu the Great might just give the idea an approving nod.

CHAPTER 5

Bank It for a Dry Day

And it never failed that during the dry years the people
forgot about the rich years, and during the wet years they
lost all memory of the dry years.
—*John Steinbeck*

No one expected water grabs like those popping up across sub-
Saharan Africa to suddenly appear in the western United States. But in
March 2014, Saudi Arabia's giant food company, Almarai, bought about
15 square miles (40 square kilometers) of farmland 100 miles (160 kilometers) west of Phoenix, Arizona. With that $47.5 million purchase
came 15 wells capable of pumping water from deep underground. The
Saudis had set up shop in Arizona, perfectly legally, to use the arid western state's precious groundwater to grow alfalfa hay. Their intent was
not to feed US livestock, but to ship the hay back home to the Middle
East to feed Saudi dairy cows.[1]

The story of how a nation half a world away ended up tapping water
from beneath the American desert begins about four decades ago. After

the oil embargo of the 1970s, the Saudis realized that they were vulnerable to a retaliatory grain embargo and determined to make their nation self-sufficient. They used vast oil revenues to heavily subsidize land, labor, equipment, and irrigation water in order to grow massive quantities of wheat in the sandy desert. From a few thousand tons in the mid-1970s, Saudi wheat production climbed to 5 million tons in 1994. Saudi Arabia not only had enough wheat for its own people, it exported the staple to other nations.[2]

But there was one big problem: water. Saudi Arabia gets only 2.3 inches (59 millimeters) of rainfall a year. Much of that moisture evaporates right back to the atmosphere. The runoff remaining after those releases to the air was far too small to achieve the nation's ambitious goals. So the House of Saud decided to go underground. Beneath the surface was a seemingly vast reservoir of water that nature had filled about 20,000 years earlier, when the region's climate was much wetter. So engineers installed deep wells, giant pumps, and circular sprinklers to draw the water out and irrigate vast areas of wheat. At the peak of production, Saudi farmers were pumping out nearly 20 billion cubic meters (16.2 million acre-feet) of water a year, a volume bigger than the annual flow of the Colorado River, and more than eight times Saudi Arabia's renewable supply.[3]

A groundwater account is much like a bank account: if withdrawals exceed deposits, the account shrinks. At some point, usually before the account registers empty, it gets too expensive or difficult to lift the water from ever-greater depths, or the deeper water becomes too salty to use. In the Saudi case, when oil revenues dropped dramatically, the government could no longer afford to prop up its costly and unsustainable wheat production. Within two years in the midnineties, Saudi grain output fell by 60 percent. Then in January 2008, with their aquifers severely depleted, the Saudis decided to phase out irrigated wheat production altogether. Instead, they would do what most countries do when they cannot produce their own food: import it. But the Saudis also started buying farmland in Ethiopia, Egypt, the Philippines, the western United States, and elsewhere so they could use other countries' water to grow their crops.[4]

If the Saudi groundwater tale were an isolated case, it might not be so troubling. But it's not. Groundwater depletion is a worldwide phenomenon that poses serious risks to food security, rural livelihoods, freshwater ecosystems, and future generations. Because underground water is out of sight, it is often out of mind as well. Rarely monitored or regulated, groundwater depletion is the sleeping tiger of global water threats.

Groundwater plays a vital role in the global hydrologic cycle. About 96 percent of all the freshwater on Earth that's not bound up in ice is stored underground. Only a tiny fraction of that groundwater, however, cycles between the sea, air, and land in a given year and is thus renewed. Aquifers provide reliable, high-quality supplies of water, and unlike reservoirs on the land, allow little water to escape through evaporation. The slow and steady discharge of water from shallow aquifers also keeps many rivers flowing through the dry season. And of growing importance, aquifers offer protection during times of drought: when rivers and surface reservoirs shrink due to lack of rain and snowmelt, farms and cities can turn to groundwater to get them through the dry spell.[5]

In the middle of the twentieth century, as access to electricity expanded to rural areas and as well-drilling and pumping technologies improved, groundwater use took off. Farmers no longer had to wait for the government to build a dam and canal system to supply them with water; with enough cash or credit, they could sink a well and start pumping. In India, for example, the area irrigated by groundwater climbed more than 100-fold between 1961 and 1985, from 100,000 to 11.3 million hectares (247,000 to 28 million acres). Because groundwater could be tapped whenever needed, crop yields often increased, boosting food production and rural incomes. Today about 40 percent of global irrigated agriculture depends on groundwater. It is vital to feed the world.[6]

But not long after the rush to drill and pump took off, signs of trouble appeared. In 1970, a well went dry in Deaf Smith County in northwest Texas. A farmer had installed it in 1936 to irrigate a parcel of the dry lands known as the High Plains. The well drew water from the Ogallala, one of the planet's great aquifers, which underlies portions of eight US states. In northwest Texas, the Ogallala gets very little recharge from rainfall today. Water had seeped into the aquifer slowly during the Ice Age tens of thousands of years ago. This "fossil" groundwater is like oil in that once pumped from the earth, it is largely gone from that place. More than three decades of pumping had caused the water level in that Deaf Smith County well to drop 24 meters (78 feet).[7]

It was a harbinger of things to come. As the region's groundwater use went up, water levels went down. The US Geological Survey (USGS) estimates that since 1940 the High Plains Aquifer, which now supplies 27 percent of US irrigated land, has lost some 266 million acre-feet (328 billion cubic meters), a volume of water equivalent to two-thirds of the water in Lake Erie. Water tables in a sizeable area of Texas have dropped 100–150 feet (30–46 meters).[8]

"The High Plains Aquifer is Nature's nearly perfect water storage system," said then USGS director Marcia McNutt upon the release of the agency's study in 2011. "In less than 100 years we are seriously depleting what took Nature more than 10,000 years to fill."[9]

Not long after the bell tolled in Deaf Smith County, Texas, pumps began to suck air in other regions vital to irrigated agriculture around the world—including the north plain of China, northwest India, Pakistan, and other areas of the western United States. In the late nineties I took a stab at estimating how much groundwater was being depleted worldwide. Using the best data I could find and making some back-of-the-envelope calculations, I estimated that the overpumping of aquifers totaled some 200 billion cubic meters (162 million acre-feet) of water per year, a volume sufficient to grow 10 percent of the world's grain supply.[10]

My estimate was crude, but, as it turns out, in the ballpark. In recent years, scientists have acquired better data and deployed more sophisticated tools to estimate the global scale of groundwater depletion. Using state-of-the-art hydrologic models along with estimates of groundwater withdrawals, a team led by Yoshihide Wada of Utrecht University in the Netherlands estimated in 2010 that annual global groundwater depletion in 2000 had totaled some 283 billion cubic meters (229 million acre-feet), roughly equal to the annual flow of 15 Colorado Rivers or 3.5 Nile Rivers. A year later, Leonard Konikow with the USGS took a different approach based on volumetric measurements rather than water-budget accounting. He arrived at an annual global depletion figure of 145 billion cubic meters (118 million acre-feet), about half as large as Wada's estimate.[11]

Konikow also compared rates of depletion by decade for the last 108 years and documented that groundwater depletion is accelerating. He found, for example, that nearly twice as much groundwater was depleted annually during the period from 2001 to 2008 (the last year in his analysis) than during 1981–90. Depletion was speeding up particularly fast in the Ogallala Aquifer: the depletion rate during 2001–8 was triple that of 1991–2000, and nearly six times greater than during the period 1981–90.

In 2002, NASA launched a mission that for the first time allowed us to "see" what was happening to water stored underground. The Gravity Recovery and Climate Experiment (GRACE) deploys a pair of satellites to measure variations in Earth's gravity over time and space. Because gravity is influenced by mass, including the weight of water, scientists can use these measurements to estimate water depletion over time. For areas of China, northern India, the Middle East, and the western United States, the picture painted by these eyes in the sky is troubling.

When Stephanie Castle of the University of California–Irvine and her colleagues analyzed GRACE data for the Colorado River basin from

December 2004 to November 2013, what they found stunned them: the basin had lost nearly 53 million acre-feet (65 billion cubic meters) of water—and 77 percent of that loss, some 41 million acre-feet (50.6 billion cubic meters), was groundwater. The basin was in the midst of its driest 14-year period in a century, and during that decade water equivalent to one-and-a-half Lake Meads had been pulled out of its aquifers. Roughly over the same period, Lake Mead itself had lost half of its capacity. But while the lake's decline was common knowledge, thanks to the famous white bathtub ring around its perimeter, the groundwater story had remained invisible—until Tom and Jerry, the nicknames given to NASA's twin satellites, revealed it.[12]

"We don't know exactly how much groundwater we have left, so we don't know when we're going to run out," Castle said. "This is a lot of water to lose. We thought that the picture could be pretty bad, but this was shocking."[13]

GRACE has delivered other unsettling news, as well. Data for northern India, often called the nation's breadbasket for its bountiful harvests of wheat and rice, showed that groundwater depletion was averaging 18 billion cubic meters (14.6 million acre-feet) a year, or a total of 109 billion cubic meters (88.4 million acre-feet) over the six years scientists analyzed, 2002–8. NASA's Matthew Rodell, who headed up this work, and his colleagues point out that annual rainfall during this period was close to normal, so presumably the depletion would be even greater during a drought. Meanwhile Jay Famiglietti, senior water scientist at the NASA Jet Propulsion Laboratory in Pasadena, California, and a professor at UC–Irvine, led the GRACE team's look at groundwater depletion in California's Central Valley, the fruit and vegetable bowl of the United States. They found that in less than a decade a volume of groundwater equal to about two-thirds the volume of Lake Mead had been depleted.[14]

Summing up a decade's worth of global findings from GRACE,

Famiglietti warns that "because the gap between supply and demand is routinely bridged with nonrenewable groundwater, even more so during drought, groundwater supplies in some major aquifers will be depleted in a matter of decades. . . . Vanishing groundwater will translate into major declines in agricultural productivity and energy production, with the potential for skyrocketing food prices and profound economic and political ramifications."[15]

The solution to the global groundwater dilemma might seem obvious: pump less and refill depleted aquifers. But both are much easier said than done. In Texas, for example, under the "rule of capture," farmers may pump as much water as they want from beneath their land. In 2011, the High Plains Water District, which is based in Lubbock and covers a 16-county area of West Texas atop the declining Ogallala Aquifer, became one of the first districts to place a cap on the volume of groundwater farmers can pump. The rule also required that new wells be equipped with meters to measure groundwater use. Some irrigators quickly threatened lawsuits over what they saw as violations of their property rights, leading the water district to delay enforcement and penalties. Then in February 2012 the Texas Supreme Court confirmed that landowners own the water beneath their property, but also acknowledged the authority of agencies like the water district to regulate groundwater.[16]

The water district's cap on pumping is a good attempt to slow the depletion. It sets pumping allowances that diminish gradually over the first few years, giving farmers time to adapt. Its overall goal is not to stop the depletion, but to ensure that at least half of the water stored in the Ogallala in 2010 will still be there in 2060—a planned depletion that would at least save some groundwater for future generations. While some irrigators are on board with the approach, others say the rule could make a decent harvest impossible, especially in a drought year. Although the rule and reporting requirements are technically in effect, as of April

2016 the district had made the metering requirement for new wells optional and was still working out the rule's enforcement provisions.[17]

Replenishing an aquifer has complications too, not least of which is finding a source of water to send underground. There is a long history of recharging groundwater around cities and towns. New York's Nassau County on Long Island began using wells to recharge its aquifers, the sole source of its drinking water, in 1937. Likewise, Orange County, California, started replenishing its groundwater supply in 1936. Today the county has one of the largest recharge systems in the world. By injecting highly treated wastewater underground, it has more than doubled the yield of its groundwater basin. And Arizona has used a portion of its allotment of the Colorado River to replenish its aquifers, effectively banking water underground for future use.[18]

But replenishing groundwater depleted by intensive, large-scale irrigation is a horse of a different color. Injecting wastewater through wells may be cost-effective for a municipal utility but is generally too expensive for a farming district. Large-scale recharge also requires "surplus" water to direct underground. With most surface water already accounted for in so many places where groundwater is being depleted, there is often little water to be found that doesn't already belong to another user, whether a city, a farming community, or a river or estuary.

Moreover, once water is pulled out of the pores of an aquifer, the rock formation can compact, similar to the way a wet sponge shrinks after it dries out. With an aquifer, this loss of pore space can become permanent, preventing it from holding water again. Compaction can also wreak havoc with roads, canals, and other infrastructure as the surface of the land sinks from the compression of the rock below. In the southern portion of California's Central Valley, land has been subsiding as much as 1 foot a year, causing water canals to crack and roads to buckle. A now famous photograph shows a researcher standing beside a tall pole southwest of Mendota, California, in 1977, with a sign 28 feet above his head demarcating where the land surface had been in 1925.[19]

But as dramatic and costly as land subsidence can be, of greatest concern is the loss of space to store water if an aquifer's depletion is not reversed soon enough. "Infrastructure can be fixed," said Michelle Sneed, a hydrologist with the US Geological Survey, speaking to the *Los Angeles Times* in 2015. "The storage capacity cannot."[20]

~

Heading south from California's capital city of Sacramento, it's hard to believe what the state's Central Valley, bordered by the Sierra Nevada to the east and the Coastal Range to the west, once was: a lush wilderness of freshwater marshes, towering oak and cottonwood forests, abundantly flowing rivers, and a rich mosaic of grasslands studded with wildflowers. It was a crucial artery along the migratory bird route known as the Pacific Flyway. About 450 miles (720 kilometers) long and consisting of the Sacramento River valley in the north and the San Joaquin River valley in the south, the Central Valley hosted unimaginable numbers and varieties of geese, ducks, swans, and songbirds. When in flight, the waterfowl would darken the sky.[21]

Today some 99 percent of the grasslands and 90 percent of the wetlands are gone. In their place is one of the most abundant and lucrative pieces of agricultural real estate in the world. Alfalfa for dairy cows, grapes for California wines, and orchards of almonds, walnuts, pistachios, and oranges now dominate the landscape. To create the bountiful harvests the nation and increasingly the world have come to depend upon, farmers tap copious quantities of the valley's water to supplement the rains. In times of drought, when the rivers fed by snowmelt flow low and surface reservoirs shrink, farmers turn to their critical backup supply: groundwater.

Just west of Modesto in the San Joaquin Valley, Nick Blom, an affable man in his forties with an easygoing manner, farms 1,300 acres (526 hectares) with his father and brother. Tidy rows of almond trees cover 700 acres (283 hectares), with smaller areas in oranges, walnuts, and wine grapes. When Blom left his graduate school experiments at California State University in Fresno 20 years ago to start farming, he had no idea he'd end up playing a part in a much larger experiment that could partially remedy California's massive groundwater problem. He is one of a small group of farmers who have volunteered to work with scientists to determine whether purposely flooding farmland with excess floodwaters from winter storms or snowmelt can help replenish the state's groundwater without either polluting it or harming the farmer's plants and crop yields.[22]

"It's a risk I'm willing to take," Blom said, standing amidst the 5 acres (2 hectares) of 15-year-old almond trees he has dedicated to the experiment.

In normal years, Blom gets nearly all the irrigation water he needs from the Modesto Irrigation District, which supplies surface water to his farm via canals. But in 2015, the fourth year of an ongoing drought, the district supplied less than half of his usual allotment. To get by, he purchased some water from neighbors and pumped groundwater to make up the difference. Along with the severe drought, something else was motivating Blom—a state law passed in 2014 that requires groundwater users to work together to slow the depletion of their aquifers and make their water use more sustainable. If they fail to organize locally, the state can step in and develop plans for them.

"I'd rather be part of decision making than have to deal with the decisions that are made," said Blom, who serves on the irrigation district's board.

In the spring of 2015, forecasters began predicting a much wetter-than-normal winter due to an unusually strong El Niño weather pattern. Scientists at the University of California–Davis seized on the likelihood of winter floodwaters materializing to get their experiments going. They set up a number of projects, including one with Blom and the irrigation district to deliver surplus storm flows to his 5 acres (2 hectares) of almond trees. Between late December and late January, when the trees are dormant, they flooded the orchard with 6–8 inches (150–200 millimeters) of water on four separate occasions. Each time it took about a day for the water to seep through the sandy-loam soils to the groundwater below.

With funding from the Almond Board of California, the UC–Davis scientists are monitoring the groundwater levels beneath Blom's farm, checking for nitrate pollution that might occur as the floodwaters pass through the fertilized soils, and studying the trees for any signs of harm. Through a plastic tube buried underground, they are photographing the almond trees' roots to detect any damage. As of 2016, there had been no obvious effects from the flooding on the almond trees' health or yield. The researchers note, however, that it might take several years for environmental stresses to make their mark. The UC–Davis team has already studied off-season groundwater recharge on alfalfa, grapes, and pistachios, and has so far found no harm to the crops from winter flooding of fields.[23]

Helen Dahlke, a UC–Davis hydrologist, says the team hopes to execute three years of treatment and monitoring at Blom's farm. Even with a control plot for comparison, it takes time and experimentation to understand how the almond trees respond to the flooding. Compared with annuals, such as alfalfa, "trees are definitely more complicated," Dahlke said.[24]

A lot is riding on the outcome. The Central Valley's warmth, sunshine, and soils make it a prime location for growing high-value fruits and nuts. But every almond or berry requires water, and with growing competition from both urban and environmental uses, that water has increasingly come from beneath the earth. Statewide, 2 million acre-feet (2.5 billion cubic meters) of groundwater is being depleted per year, with the majority of those losses occurring in the Central Valley. During the severe drought of 2012–16, water levels went into free fall as farmers

An almond farm in California's Central Valley is flooded in the wintertime to replenish groundwater supplies while scientists study the effects on soils, tree health, and water quality. Photo by Joe Proudman/UC Davis.

pumped more to offset reductions in their surface water deliveries. In 2015, farmers were expected to pump an additional 6 million acre-feet (7.4 billion cubic meters) to partially fill a surface water shortage of 8.7 million acre-feet (10.7 billion cubic meters).[25]

In contrast to annual crops, which farmers can choose not to plant during times of drought, orchard trees are a long-term investment and need water to stay alive. California's almond trees, which typically live about 25 years, require some 4 feet (1.2 meters) of water per year. Even in good times, rainfall provides only slightly more than a quarter of that demand; the rest must come through irrigation. During the recent drought, many almond growers pumped much more groundwater than usual to keep their yields up and their trees alive. And, whereas rice, hay, and other thirsty crops might grow just as well in regions where water is more abundant, almonds are quite particular about their grow-ing environment. Thanks to its Mediterranean climate, California is one of the best almond-growing regions in the world. "They have to be grown here," said Gabriele Ludwig, the Almond Board's director of sus-tainability and environmental affairs. "This is a prime almond-growing area."[26]

To meet the rising demand for the tasty nut, as well as the butter and milk produced from it, California farmers have roughly doubled the area planted in almond trees over the last 20 years. Today Califor-nia's roughly 1 million acres of almond orchards produce some 2 billion pounds of almonds. Two-thirds are exported to other countries. With such large areas and sales at stake, and big cutbacks of surface water dur-ing the drought, the almond industry is worried about the requirements of the new groundwater act.

"You have decades of livelihoods based on access to groundwater," said Ludwig. "With tree crops you have a tremendous investment up front. Sudden changes in regulations don't work well with trees." As for the new groundwater act, "It's scaring the living daylights out of us."

Yet most agree that better groundwater management is crucial to the long-term future of farming. That's why Ludwig and the Almond Board are supporting recharge experiments on almond orchards like those at Blom's farm. And while California has justifiably been viewed as negligent for failing to regulate and manage its groundwater for so long—consider that Arizona passed its pioneering groundwater law in 1980—the act signed by Governor Jerry Brown in September 2014 stands a fighting chance of significantly slowing, if not reversing, the draining of the state's groundwater accounts.

The Sustainable Groundwater Management Act (SGMA) puts forth a three-phased timeline. By 2017, California's stressed groundwater basins, which encompass about 96 percent of the state's average annual groundwater supply, must establish a local groundwater sustainability agency that has the authority to tax, regulate, and limit pumping. Basins covered by the act then have until 2020 (or 2022 for the less-critical ones) to develop their plans. Each basin must formulate a water budget, and set goals to avoid a half dozen undesirable consequences: the chronic lowering of groundwater levels, the reduction of groundwater storage, the degradation of water quality, seawater intrusion, land subsidence, and the depletion of rivers connected to groundwater. The basins have until 2040 to actually achieve these goals, but they must report on their progress every five years. If state officials believe implementation is too slow, they can step in and assume control.

"That's the stick," said Andrew Fahlund, a senior program officer at the California Water Foundation in Sacramento, who helped write the law. "The feeling is, 'we don't want the big bad state (water) board coming in and taking over.'"[27]

But the choice to keep decision-making and planning local is an important hallmark of the California law. Voters approved a state bond to generate $100 million in funding to help the local groundwater basins, many of which will need outside technical expertise, develop their plans. The Farm Bureau, which had opposed passage of the groundwater act, is now on board with helping to implement it.[28]

No one, however, thinks the road ahead will be easy. "This is a problem that's been a hundred years in the making," Fahlund said. "Fixing it will involve some pain for the California economy and society."

Much will change between now and 2040 as basins work to balance their water budgets and achieve their sustainability goals. Faced with new fees and pumping limits, farmers will variously switch to less-thirsty crops, periodically fallow some land, upgrade to more-efficient irrigation systems, monitor their soil moisture, better schedule their irrigations, and generally figure out ways to maximize the value of their more-limited groundwater supply. They will likely do more trading of water among themselves and with other water users. Inevitably, they will take some cropland—perhaps as much as 15–20 percent in the southern portion of the Central Valley—out of production altogether.

While by no means a silver bullet, active replenishment of depleted aquifers can certainly help. A 2015 study of the eastern San Joaquin Valley matched up suitable soil types, winter water availability, and infrastructure, and concluded that an annual average of 80,000–130,000 acre-feet (98–160 million cubic meters) of surplus winter flows could be available for recharge. About 55 percent of that water would boost underground water storage, and 43 percent would sustain base flows, making the river healthier. Altogether such a plan could slow depletion in that part of the valley by an estimated 12–20 percent.[29]

Groundwater recharge is also a more cost-effective way to increase water storage capacity in California than building new dams and surface reservoirs. Increasing groundwater storage through winter flooding of farm fields, like the project at Nick Blom's farm, typically costs $40–$107 per acre-foot, according to Daniel Mountjoy, director of resource stewardship with Sustainable Conservation, a nonprofit organization based in San Francisco that is coordinating recharge experiments throughout the state. Using dedicated recharge basins, where no crops are grown, costs $145–$200 per acre-foot, depending on the cost of the land. By contrast, the proposed Temperance Flat Dam on the upper San Joaquin River is estimated to cost more than $1,500 per acre-foot, 7–37 times more than groundwater replenishment.[30]

In partnership with scientists and farmers, Sustainable Conservation has identified 130 different fields planted in 10 different crops spanning a total of 14,000 acres (5,670 hectares) for possible testing of on-farm recharge. These sites include Blom's almond orchard and the other projects overseen by Dahlke and her UC–Davis team.[31]

In early 2017, the experimental work, irrigators-at-large, and depleted aquifers all got a boost from the intense precipitation that hit the state. With rivers running high and engineers opening dams to make room in surface reservoirs to capture more storm runoff, there was ample floodwater available to direct underground. Madera, Fresno, and Tulare Irrigation Districts were among those offering water to their growers to flood their fields and let the water infiltrate to the aquifers below. The Madera District routed water it received from releases from Friant Dam to designated recharge basins. It also offered water to its landowners free for about 10 days, after which it charged a reduced rate. "The District recognizes the need to capture and use as much surface water as possible when it is present," it said in a press release that urged its irrigators to sign up to take water. In a similar vein, the Fresno Irrigation District's general manager, Gary Serrato, told the *Fresno Bee*, "We are trying to capture everything we can for recharge."[32]

Support for replenishing aquifers is also growing among crop associations and food and beverage companies. In addition to funding scientists to study particular sites, like Blom's farm, the Almond Board is funding work to identify which of the vast acreages of almond production are potentially suitable for groundwater recharge. WhiteWave Foods, based in Denver, Colorado, is supporting groundwater recharge projects to help its suppliers become better water managers and to balance the water footprint of the company's almond milk and other products.[33]

"All are reading the tea leaves and seeing that the alternative is to stop pumping," Mountjoy said.

While it is too soon to tell, California's new groundwater law may just unleash the broader reforms the state needs to maximize its limited water supply. Farmers who forgo crop production may eventually be compensated not only for replenishing groundwater but also for restoring bird and wildlife habitat in the Central Valley. More broadly, the law might force communities to recognize the long-denied hydrologic truth that groundwater and surface water are connected and need to be managed together, as one pool rather than two. And the collaborative work needed to conserve groundwater could turn a me-first philosophy into one that better serves the community as a whole.

As Blom, the almond grower, said, "It's not just our groundwater. It's our neighbors' to the north and south too."

~

No country has a greater stake in solving the problem of groundwater depletion than India, the world's second-most-populous nation. About 120 million Indians, one-quarter of the workforce, are farmers. Access to groundwater is a cornerstone not only of the nation's food security but also of rural livelihoods and social cohesion.

Back in the 1960s, the Indian government pushed forward a "green revolution" in agriculture to stave off famine and build food security. Along with fertilizer and new crop varieties, water was a key ingredient. As the government built dams and canals to deliver surface water for irrigation, farmers were encouraged to construct boreholes and dig wells to put groundwater to use. Because farmers could pump groundwater when they needed it, its use often boosted harvests far above those dependent on water deliveries from large centralized canal projects. Although hunger and malnourishment still stalk India's poorest people, national food production climbed—and the tapping of groundwater was a key driver of this success. Today India pulls twice as much water from beneath the earth as China does, and 89 percent of it is used to irrigate crops. Some 25 million wells dot India's countryside. Two-thirds of them were sunk after 1990, and most belong to small farmers.[34]

Five decades of explosive growth in the digging and drilling of individual wells has created what economist Tushaar Shah, a senior fellow with the International Water Management Institute (IWMI) in Colombo, Sri Lanka, and an expert on South Asian water policies, calls a mounting "anarchy." The government often charges a flat fee for electricity, a subsidy that can benefit the rural poor, but that allows irrigators to pump as much water as they'd like for the same price. So instead of regulating groundwater, the government is often encouraging excessive pumping. India's widespread groundwater depletion, writes Shah, who is based in the western Indian state of Gujarat, "raises questions about the future of a vast agrarian system founded on a boom that seems destined to go bust."[35]

But innovative efforts at the village and state levels to battle the depletion problem have generated some hope. Taken together, they show that moving toward sustainable use of groundwater is as much about incentives and governance as it is about hydrology and technology. Success, however, will not come easily.

In northeastern Rajasthan, a western state in the Indian peninsula, the Alwar District sits at the northern edge of the ancient Aravalli mountain range and covers about 8,380 square kilometers (3,235 square miles). Its roughly 2 million people are spread among five towns and 1,991 villages. For their livelihoods, they grow wheat, beans, and other staple crops on small farms of no more than 2.5 hectares (6 acres), and raise buffalo, camels, cattle, sheep, and goats. The semiarid region receives only 350–400 millimeters (14–16 inches) of rainfall a year, and the vast majority of it comes during the three to four months of the monsoon season, typically June to September. The rest of the year is dry.[36]

For centuries prior to the arrival of the British in 1600, most villages in peninsular India collected monsoon rains and runoff in tanks for use during the dry season. The tanks also helped control flooding and recharged groundwater. But with colonial and then state control over water and forested catchments, many of these traditional village structures fell into disrepair. In Alwar, the logging of trees caused erosion, silting up the tanks. With the loss of their water-harvesting infrastructure, the villagers suffered from lack of water for their crops and livestock, especially during droughts. The Alwar District, once known locally as *naha* (an area with a high water table), saw its groundwater level sink. Crop yields, food security, and village well-being sank along with it.[37]

Into this picture strode Rajendra Singh, then a 28-year-old development worker trained in traditional Ayurvedic medicine and planning to be a doctor. Originally from the state of Uttar Pradesh, he had moved to Jaipur, the capital of Rajasthan, and in 1984 cofounded an organization called Tarun Bharat Sangh (India Youth Association). TBS followed a Gandhian philosophy of helping the poor help themselves. In 1985 Singh traveled to the Alwar District with the aim of introducing health and education programs. But he quickly learned that something else was more important to the villagers than schools and hospitals: water. With an ongoing two-year drought, women were walking 2 kilometers (roughly 1 mile) a day to collect water rationed from wells. Animals were dying and people were malnourished. Men were leaving the village to look for work in the cities.

With the help of local villagers and advice from an elder, Singh and the TBS team built three *johads* (earthen water-harvesting structures) patterned after those traditionally used for centuries. About 4.6 meters (15 feet) deep, the structures captured rainfall and recharged the groundwater. Dry wells began to refill. To construct more *johads*, TBS paid villagers in wheat for their labor. The villages established assemblies to select the catchment areas and project sites, and to create budgets. After construction of the *johads*, groundwater recharge grew by 20 percent. Soil moisture increased as well, making crops more resilient during the dry season. About one-fifth of local runoff was better spread out over the year, which revived perennial flows of the Arvari River.

As water levels rose in more wells and word spread to more villages, water harvesting became something of a mass movement. Today, three decades after Singh and TBS arrived in Rajasthan, 10,000 *johads* populate the landscape and 1,000 villages again have a reliable supply of water. With the rise in groundwater levels, several dried-up rivers have begun to flow again.[38]

"This change which has been wrought in our region is a change toward prosperity," said Singh, who won the prestigious Stockholm Water Prize in 2015 and is now dubbed the "water man of India." An area that was once "so barren and dry, is now green and fertile again. This work of ours is a way to solve both floods and droughts."[39]

The successes of Singh and TBS in northeastern Rajasthan have inspired others in the region as well. India's branch of SabMiller, which works globally to lessen the water footprint of its beer-making operations, has partnered with farmers and other stakeholders in the vicinity of Neemrana, a historic town in the Alwar District and the location of its Roches Brewery, to reduce the risks of groundwater depletion. A hydrologic study had found the region's water table to be dropping 0.9 meter (3 feet) per year. The company supported the construction of six rainwater-harvesting structures near the brewery's groundwater wells, sizing them to recharge at least as much water as the brewery withdraws. The company also helped some 4,000 farmers adopt more water-efficient practices, which boosted their productivity and net incomes by more than 20 percent. With more recharge and fewer withdrawals, the water table rose some 5.2 meters (17 feet). The basin's groundwater use is not yet in balance with recharge, but the gap has narrowed by nearly one-third.[40]

In the Indian state of Gujarat, which borders Rajasthan and has also seen large-scale depletion of its aquifers, a couple of innovative and promising ideas are being piloted. IWMI is promoting the use of solar-powered groundwater pumps with a buyback guarantee for any surplus power produced. The buyback of excess electricity gives farmers an incentive to pump only the water they really need for their crops, which curbs the depletion of groundwater while diversifying farmers' incomes. And by pumping with solar energy rather than diesel fuel, they reduce climate-altering carbon emissions. IWMI made its first payment in June 2015 to a banana and wheat grower for his "solar crop." It's an innovative solution that confronts groundwater depletion, climate change, and rural poverty all at once.[41]

The Gujarat government itself is pioneering a novel approach to balancing its groundwater accounts through the rationing of electricity. The government spent $250 million to put 1.1 million irrigation wells on separate electricity feeders that allow it to limit power to the wells to eight hours per day. This effectively caps the amount of groundwater a farmer can withdraw. It also creates, in the words of Tushaar Shah, the economist with IWMI, a "switch on–switch off groundwater economy," whereby the government can help farmers weather a dry spell by increasing the power, and then, when rains are good, curtail pumping by limiting power. So far, the scheme is working. Whereas water levels in large areas had previously declined even during the monsoon season, now as a rule they rise. Gujarat's agricultural economy continued to grow at a rapid clip, and, according to Shah, Gujarat is the only state in western India where groundwater levels are improving.[42]

The ongoing efforts to replenish aquifers in India, the western United States, and elsewhere are promising. But ending the depletion of groundwater, which ranks among the gravest threats to future food security, will require major reforms in government laws and policies. It is troubling enough to leave future generations with a less stable climate. It is a form of intergenerational theft to deplete the planet's underground water reserves as well.

CHAPTER 6

Fill the Earth

You'll be the mule, I'll be the plow
Come harvest time we'll work it out
There's still a lot of love here in these troubled fields.
    —*Nanci Griffith*

NANCI GRIFFITH'S "TROUBLE IN THE FIELDS" is one of my all-time-favorite songs. With a haunting melody and poignant lyrics, it speaks of love of place, and of the hard work and perseverance that farm life requires. It also speaks of an event in US history as dark as the skies during its worst days: the Dust Bowl.

The plowing up of native prairie grasses and their replacement with cultivated wheat early in the twentieth century turned the Great Plains into a breadbasket for the nation. But when a multiyear drought during the mid-1930s coincided with vicious westerly winds on the panhandles of Texas and Oklahoma and nearby areas of Kansas, Colorado, and New Mexico, this manmade landscape literally blew away. "Black blizzards" choked people and animals. Storms of dust rained down on

Chicago; Washington, DC; and even ships in the Atlantic Ocean. By 1940 more than 2.5 million people had fled the region. The Dust Bowl was Nature's wake-up call, a bitter reminder of the workings of the land and our relationship to it. Eighty years later, as we face climate change and water shortages, it is past time to heed that call.[1]

When we hear someone described as a "water manager" we most likely imagine that person operating a dam, running a treatment plant, or piping drinking water to a city or town. Rarely do we conjure up a farmer or rancher. But those who work the land for a living manage what is arguably the most underappreciated yet vital component of the water cycle—that precious band of Earth we call soil.

The world's soils hold about eight times as much water as all rivers combined. This soil reservoir is the principal water supply for forests, rangelands, and croplands; how much water it contains will dramatically influence our food security in the coming decades. On average our diets require about 1 liter of water per calorie. Plants lift water from the soil and use it in the miraculous process of photosynthesis, whereby sunlight converts carbon dioxide and water to carbohydrates and oxygen. Photosynthesis is the foundation not only of global food production but also of terrestrial life—and without sufficient water in plants' roots, it cannot occur.[2]

Today, a quiet revolution is taking place in our understanding of that soil reservoir and how land management influences the volume of water it holds. A growing number of scientists, research institutions, government agencies, farmers, and ranchers are realizing that to feed 9 billion people, we must farm and graze livestock in a way that regenerates soils. New methods, if taken to scale, can also sequester vast amounts of carbon in the soil, fighting climate change even as they build resilience to its effects.

One farmer at the vanguard is Gabe Brown, a middle-aged, tell-it-like-it-is North Dakotan who farms some 5,400 acres (2,200 hectares) with his family on the outskirts of Bismarck. Brown and his wife, Shelly, bought the farm from Shelly's parents in 1991 and initially farmed it as most midwesterners do: they planted a cash crop (in their case wheat), and after the harvest they left the field fallow until the next planting. They tilled the soil and added chemical fertilizer for nutrients. Their livestock grazed the native rangeland all season long. Two years in, they realized that with only 16 inches (406 millimeters) of annual precipitation, they needed to conserve more moisture in the soil. So they stopped turning the soil over, which had been allowing precious moisture to evaporate, and they bought a no-till drill to implant seeds directly into the soil. But four consecutive crop failures due to hail or drought during the midnineties pushed them to make more radical changes.[3]

"It's all about soil health," Brown said to a rapt audience at the annual conference of the Quivira Coalition, a nonprofit dedicated to expanding the "radical center" where progressive farmers, ranchers, and environmentalists find common ground. This 2012 conference in Albuquerque, New Mexico, was focused on the key elements of regenerative agriculture. I was invited to speak about water, and Brown to address soil. At that gathering I realized how blinkered I myself had been about agricultural water use, despite having written on the topic for many years. I had researched how to get more "crop per drop" through efficient and judicious irrigation, and practices like terracing and water harvesting that direct more rainfall into the soil. What I had neglected, though, was the importance of repairing the soil reservoir itself—the living matrix that holds water in its pores and absorbs it like a sponge. And that's what Brown, whose farm has become a nationally recognized test case for holistic and regenerative agriculture, was talking about.

"There's no place in production agriculture for tillage because you're just destroying the home of all that soil biology," Brown said. "The soil is alive. Instead of focusing on the plant, we have to focus on the life in the soil."[4]

That "soil biology" Brown refers to is the community of bacteria, fungi, protozoa, mites, and worms that in large part determines the health of the soil, which in turn determines how rapidly water seeps into it and how effectively it stores water. A teaspoon of healthy soil can harbor as many as 1 to 7 billion of these hardworking critters. They form the soil into clumps that adsorb water and open up pores that provide air and hold moisture. If the soil surface is compacted, water has nowhere to go and will simply pool and evaporate, or run off. But soils with a healthy microbial community and a good amount of organic matter allow rainwater to rapidly infiltrate and hold it in the root zone for use by plants.

Any particular soil's ability to retain water is a complex function of its structure and composition. The role organic matter plays, and specifically organic carbon, can vary significantly with the type of soil. A team of scientists with the US Department of Agriculture found, for example, that when starting from low levels of organic carbon, a rise in carbon increases water retention in coarse soils but decreases it in fine-textured soils. At higher levels of organic carbon, however, virtually all soils are able to hold more water.[5]

Humberto Blanco, a professor of soil science at the University of Nebraska–Lincoln, and his colleagues report that for medium-textured soils an increase of 1 percentage point in organic carbon can boost the volume of water in the top 20 centimeters (8 inches) of soil by an average of 12.5 millimeters (half an inch). That may not seem like much, but it adds up quickly—to 13,600 gallons per acre. Similarly, Australian scientist Christine Jones estimates that a 1-percentage-point increase in organic carbon allows the top 30 centimeters (or 1 foot) of soil to hold 168,000 more liters of water per hectare (18,000 gallons per acre). That's significant because organic carbon levels in many cropland soils have fallen 3 percentage points or more since the beginning of European settlement. Returning soils to healthier conditions could increase water by 500,000 liters per hectare (54,000 gallons per acre).[6]

The kind of industrialized monoculture farming that has come to dominate world agriculture since the middle of the twentieth century has focused singularly on increasing cash-crop yields, especially of wheat, corn, rice, and soybeans. While it succeeded in boosting food production alongside population growth and spared large areas of forest and grasslands from the plow, it paid little attention to the health of the soil. Heavy plowing exposes the soil's organic matter, releasing carbon into the air, where it contributes to global warming. It also allows more water to evaporate, depleting the soil reservoir. Without plants to root the soil in place, land erodes, nutrients disappear, and still more carbon escapes to the atmosphere. Scientists estimate that since the start of tillage-based farming, most agricultural soils have lost 30 to 75 percent of their organic carbon content, and that those losses accelerated with the expansion of industrial agriculture.[7]

To drive home the difference in soil health among different cropping systems, and to have a little fun, researchers at South Dakota State University Extension orchestrated a "tighty whities" demonstration during a 2016 soil health school for local farmers. Five weeks before the program began, five white briefs (yes, men's undergarments) were buried in the soil at each of three different farm sites: corn with conventional tillage, soybeans with mulch/reduced tillage, and a no-till field with cover crops. Since soil microorganisms require carbon to thrive, and cotton contains high levels of carbon, the state of the briefs after five weeks underground would illustrate how active the invisible critters had been. The results were indeed "revealing."[8]

The briefs from the conventionally tilled cornfield held up the best, indicating the least microbial activity. Compared with a control brief that wasn't buried, they had lost about 13.2 percent of their weight. The reduced-till field produced briefs a bit more tattered and that weighed on average 17.4 percent less than the control pair. The briefs that emerged from the no-till, cover-cropped field, however, were barely recognizable. They had lost an average of 51.4 percent of their weight. The microorganisms had left the elastic waistband alone, but had demolished the rest.

The quirky experiment clearly showed why Gabe Brown pays so much attention to the hardworking microbial herd belowground. After going no-till, Brown planted a couple of different cover crops to protect the soil biology and to add more carbon and organic matter to the soil. About a decade ago, he moved toward a much more diverse mix of warm- and cool-season grasses and broadleaves—what's come to be called a cover crop cocktail. He first learned about this approach from a Brazilian crop specialist in 2006 while attending a no-till conference in Kansas. The idea is that the diversity of cover crops aboveground encourages diversity of organisms belowground, which in turn supplies crops with the array of nutrients they need to thrive.

It's a symbiotic two-way street that mimics the native prairie. Legumes such as hairy vetch and red clover add nitrogen to the soil, virtually eliminating Brown's need for chemical fertilizer. And the continuous plant cover keeps the soil, nitrogen, and moisture in place, while sequestering more carbon. Brown's goal is to keep a living root in the ground at all times. "Something we never ever want to see is bare soil," he said.

Only a handful of the two dozen or so crops Brown now grows on his farm are grains taken to market. Most are part of the cover crop mix that feeds the microbial community and improves the health of his soil. At some point Brown began growing cover crops not just between harvests but also alongside his cash crops. He also started "mob-grazing" his livestock. Instead of allowing his animals to openly graze the range all season, he kept them bunched together and moved them frequently. He allowed them to eat about one-third of the cover crop on his fields, leaving two-thirds to feed the soil and reduce evaporation. The cattle's manure fertilizes the field and their hooves trample the crop residue into the soil. Again, Brown was mimicking nature and the way bison historically made the range more productive and diverse.

This regenerative system is night-and-day different from the energy-intensive, high-input production system Brown started with and that most farmers in the American heartland still use. But it has paid off handsomely. Before he started the transition, the level of organic matter in his fields was less than 2 percent; now it is typically above 4 percent. The rate at which water infiltrates the soil has risen 16-fold, from about half an inch per hour to 8 inches per hour. Brown scarcely uses any synthetic fertilizer, and herbicide use is down by more than 75 percent.[9]

Brown's shift has been good not only for the land but also for the consumers of his crops and his bottom line. His plants are more nutrient dense and therefore more healthful. Because the diverse mix of cover crops feeds the soil biology (the underground "herd") as well as his livestock, it has cut costs while providing multiple benefits. Brown's corn yields average 27 percent higher than the county average, while his net costs per acre are two-thirds lower. And with his son and farming partner Paul, he's still experimenting, adapting, and building resilience into his production system. He believes the soil organic matter on the family's farm could climb as high as 6 or 7 percent—close to that of North Dakota's native prairie prior to cultivation. That will allow the soil to absorb rainwater even more quickly and to hold it for a longer time, a crucial hedge against drought. That's a happy prospect in a region where both dry spells and storms may intensify with climate change.[10]

The North Dakota county of Burleigh, where Brown farms, has become something of a hotbed of regenerative agriculture. Along with Brown, several dozen other farmers and ranchers are actively working to build healthy soils by attending to the organisms belowground as much as the ones above it, all while boosting yields and profits. They've had the benefit of support from progressive scientists and conservation specialists with the US Department of Agriculture, who see their pioneering efforts as crucial to farming's future.

"If it works in North Dakota," Brown says, "it will work about anywhere."

Which got me wondering. Just how much extra water and carbon might be stored in croplands if more farmers adopted a strategy of no-tillage and cover cropping? How much could we reduce flooding, dead zones, and the effects of drought? With more carbon sequestered in soils, might we slow climate change, and thereby repair the water cycle writ large?

It's impossible to answer these questions precisely without a lot more research, but the benefits would be substantial. Cover cropping even with two or three species can reduce erosion by 90 percent and fertilizer runoff by 50 percent. And benefits would grow quickly because we'd be starting nearly from scratch. The most recent census of US agriculture—and the first to query farmers about their use of cover crops—found that only 2.6 percent of the 390 million acres (158 million hectares) of farmland sowed with crops nationwide were cover-cropped during or between harvests.[11]

That's an astonishingly low figure given the many benefits of cover cropping, but that percentage may soon start to rise. In 2016 the USDA's Natural Resources Conservation Service (NRCS) launched a new Soil Health Division to offer technical advice, financial incentives, and tools to help farmers shift toward no-till, cover cropping, and other practices that build soil. NRCS also plans to broaden its assessment of soil health to include water retention, microbial activity, and other important biophysical properties.[12]

For farmers reluctant to go whole hog the way Gabe Brown did, even modest changes can make a significant difference. Eliminating tillage and leaving crop residues on the field, for example, can reduce water losses from evaporation by 50–80 percent, says Jerry Hatfield, director of the USDA's National Laboratory for Agriculture and the Environment in Ames, Iowa. In his talk at the 2016 joint meeting of the science societies for agronomy, crops, and soils in Phoenix, Arizona, Hatfield stressed that more water in the soil means more transpiration and thus higher yields. "We're losing 20 percent of our crop 80 percent of the time due to temporary water shortage," he said with the conviction of a scientist grounded in years of experience.

Hatfield explained that crop residues like standing corn stubble create a microclimate near the soil surface by reducing bare ground and altering wind speed. When the ground is left bare, "we cook the biology out of the soil surface," which makes those soils less effective at absorbing and holding water. What matters, he continued, "is not how much rain you get in the rain gauge, it's how much you get in the soil." He pointed out that in 2012, when much of US farm country was in severe drought, producers that had reduced evaporative water losses from their soils reaped higher yields.

Globally, researchers estimate that conservation agriculture—a suite of methods that includes minimum or no tillage, cover cropping, the retention of crop residues on the land, crop diversification, the mixed planting of trees and crops (called agroforestry), and the integration of livestock with crop production—is now used on some 105 million hectares (260 million acres), about 7 percent of global cropland. While this, too, is a small share, scientists and development specialists are actively working to expand these practices, including in the world's impoverished rural areas. For poor farm families, replenishment of the soil reservoir beneath their lands can mean the difference between hunger and a satisfied belly, and between abject poverty and enough income to send their children to school.[13]

This is particularly true in sub-Saharan Africa, where some 200 million people—about one in four—are malnourished. Today only 4 percent of the cultivated land in sub-Saharan Africa is equipped for irrigation, compared with 37 percent in Asia and 18 percent in the world as a whole. Moreover, most of that irrigated land is in just four countries— Madagascar, Nigeria, South Africa, and Sudan. With all but a small share of food production dependent on rainfall alone, large swathes of sub-Saharan Africa are prone to chronically low yields, harvest failures, and famine.[14]

The silver lining is that there is great potential to get more "crop per drop" by channeling rainfall into the root zone and improving the soil's ability to hold moisture. When farmers see results from cover cropping, land terracing, and other sustainable practices, they gain confidence to invest in fertilizers and better seeds, as well as higher-value vegetable crops to take to market. It is, for many, a pathway out of poverty.

In southern Africa, for example, conservation agriculture is leading to better water infiltration; less runoff, erosion, and evaporation; and more organic matter in the soil—the same advantages Gabe Brown is realizing in North Dakota. With the addition of water harvesting—techniques that slow rainfall runoff so it seeps into the soil or that capture rain in small ponds—the benefits multiply. One study in Zimbabwe found that conservation agriculture combined with water harvesting increased farmers' gross margins per hectare four to seven times, with the greatest gains occurring in low-rainfall areas.[15]

While there are no silver bullets, building soil health is about as good as it gets. What other activity simultaneously produces more food, reduces the impacts of floods and droughts, improves water quality, shrinks dead zones, and mitigates climate change? Some quick math suggests that storing one more inch of water in the world's farmland soils would fill the soil reservoir with an additional 100 trillion gallons—a volume equivalent to two-thirds of the Mississippi River's annual flow.

Maybe that's a challenge that farmers, researchers, and conservationists could rally around.

~

Few animals get as bad a rap these days as cattle do. They are blamed for soil erosion, water depletion, overgrazed rangelands, greenhouse gas emissions, and, when eaten, human heart disease. Writing in *Slate* magazine in 2013, James E. McWilliams, a professor of history at Texas State University, flatly concludes, "There's no such thing as a beef-eating environmentalist."[16]

Often missing from such indictments of the mooing, tail-wagging, and, yes, methane-emitting bovine, however, is our role. How we choose to manage cattle determines their environmental impact, not the animals themselves. In fact, a growing body of scientific research suggests that when managed to do so, cattle can actually build soil health, improve water infiltration, enhance habitat for birds and wildlife, and simultaneously mitigate climate change. In our quest for an ethical meal—and I write this having spent a dozen years of my adult life avoiding meat—cattle might deserve a closer look.

First, cattle possess the remarkable ability to turn grass, which we and most other mammals cannot digest, into meat and milk. Without cattle, goats, and other ruminants, the world's rangelands, which cover about 40 percent of the land on Earth (excluding Antarctica and Greenland), would contribute little to the global food supply.

Second, beef is not the water guzzler it's been made out to be. In his influential 1987 book *Diet for a New America*, John Robbins wrote that producing a single pound of meat requires an average of 2,500 gallons (9,460 liters) of water. That's about 625 gallons (2,370 liters) for a quarter-pounder, not counting the bun or the ketchup. Some years later, the Water Footprint Network (WFN), a research organization based in the Netherlands, began publishing the most consistent set of global data available on the volume of water it takes to make various products, from beef and butter to coffee and chocolate. To determine each product's global average "water footprint," WFN adds together green water, which is essentially rainwater; blue water, which is the volume taken from rivers and aquifers for irrigation; and, in more recent analyses, gray water, which is the volume needed to dilute the pollution from making that product. In 2004, WFN researchers concluded that it takes on average 634 gallons (2,400 liters) to produce one beef burger, closely in line with Robbins's figure. Many researchers, including myself, have used this figure in their work.[17]

But here's the glitch. According to WFN, 94 percent of the water consumed in producing beef is green water—rain falling from the sky. Should that rainwater be counted in beef's water footprint? Some of it, yes, but much of it, no—and here's why. For products made from wheat, corn, soy, or any crop grown where a forest or grassland used to be, it makes sense to include the rainwater that grows them in their water footprint because they are now appropriating water that had previously sustained a natural ecosystem, and that ecosystem had delivered many important services, from climate mitigation to biodiversity conservation. For example, rainwater that used to sustain a tropical forest in Brazil is now watering a soybean field. And in the US heartland, corn, soy, and wheat are now consuming rainwater that previously sustained native prairie. That's an appropriation of rainwater from nature that should be counted in a crop's water footprint.

Beef is different. Most beef cattle spend the majority of their lives on natural rangeland. They eat native grasses that the rain would nourish whether the animals were grazing that land or not. Many go to feedlots for their final weeks to get fattened before slaughter—and the water used to grow the corn and other feeds they eat while in confinement should be counted as part of their water needs. Other cattle are raised or "finished" on irrigated pasture, and the water consumed to maintain that pasture should be counted as well. But unless the cattle completely destroy the rangeland (and overgrazing is a problem, to be sure), including in their water footprint the rainwater that grows the native grasses they eat for the majority of their lives makes no sense.

I have yet to see any analysis of the water consumed in beef production that takes these important distinctions into account, but there is at least one that comes close. Back in the early 1990s, two researchers in the Department of Animal Science at the University of California–Davis undertook a very detailed analysis of the water required to produce US beef. They quite meticulously estimated the volume of water the nation's beef cattle drank, the water used to grow irrigated pasture, and the water used to process the animals for marketing. While they excluded the water consumed in growing feed crops, which I think should be included, they did not subtract the irrigation runoff that makes its way back to rivers or aquifers, somewhat balancing things out. Their bottom line: in the United States, it takes 973 gallons (3,682 liters) of water to produce 1 kilogram (roughly 2 pounds) of boneless beef. That's 441 gallons (1,670 liters) per pound of beef, or about 110 gallons (416 liters) for a quarter-pounder. That's less than one-fifth the figures from WFN and Robbins, and roughly the volume of water WFN says it takes to produce two eggs or three cups of coffee.[18]

But beef's water story gets more interesting. If raised on natural rangelands through a method variously known as holistic management, rotational grazing, or managed intensive grazing, cattle can be a net benefit to the water cycle, as well as to landscape health, bird and wildlife habitat, and climate stability. The technique essentially mimics the historical behavior of bison and other herd animals on the world's grasslands. Typically, the animals roamed together in large groups, and to guard against predators they did not stay very long in one place. Their manure was a natural fertilizer, returning nutrients to the land. Their hooves pushed those nutrients into the soil and broke up the surface so rainwater could infiltrate deep into the earth. After the bison moved on, the grasses quickly regrew. The grasslands of the US Great Plains and the vast African savannahs essentially coevolved with herd animals; they depended on one another.[19]

Today, however, beef cattle are often associated with overgrazed rangelands and ugly feedlots. "Ninety percent of people think cattle are bad," said Robert Potts, president of the Texas-based Dixon Water Foundation. "But grasslands need well-managed grazing to stay healthy. We need to educate people about that."

Potts is on a mission to do that educating, as well as to advance grazing techniques that will benefit both the watersheds of Texas and the bottom lines of ranchers. Trained as a lawyer, Potts is an integrative thinker. I first met him in 2005 when he was the general manager of the Edwards Aquifer Authority in San Antonio, Texas, and he and I had been invited to speak on a panel at the University of North Texas with T. Boone Pickens, the oil tycoon and would-be water privateer. Previously Potts had directed the Texas chapter of The Nature Conservancy. In 2007 he was tapped to head the Dixon Water Foundation, which had been newly endowed with $50 million after the death of Roger Dixon, an investor in oil and real estate who believed freshwater was the next big issue for Texas.[20]

"As a relatively small foundation, we felt a need to focus," Potts said, during my visit to the foundation's West Texas operations in late October 2016. "The piece that we felt was not covered was the health of rivers and creeks, which really goes back to land management. In Texas that means ranches—and how to use livestock to improve watershed health."

The foundation operates seven working cattle ranches, four in North Texas and three in West Texas. The Mimms Ranch, so-named for the family that settled it over a century ago, spans 11,000 acres (4,450 hectares) outside of Marfa, the far-west Texas town famous in the art world as a hub of minimalist expression. Ecologically, the ranch sits in the northeastern corner of the Chihuahuan desert, a gorgeous landscape of high-elevation grasslands surrounded by rocky volcanic mountains. Rainfall averages about 15 inches (380 millimeters) a year, but can swing wildly from one year to the next, a pattern of extremes climate change is likely to amplify. Potts views improvements in soil health as crucial to building climate resilience. "It matters less how much rain you get and more how much rain you keep," he said.

At the Mimms Ranch, Potts and his ranch managers are running both an economic enterprise and a scientific experiment. They've divided the land into three parts: one where cattle graze continuously to replicate how most Texas ranchers operate; another where no grazing is done in order to demonstrate what happens when the land simply rests; and a third that's dedicated to rotational grazing, or what Potts prefers to call high-intensity, short-duration grazing. With electric fencing, the Dixon team has subdivided 7,500 acres (3,000 hectares) into 30 paddocks, each of which is then partitioned with moveable fencing into 12 subsections. This creates 360 grazing zones, allowing the cattle to be moved to a new area just about every day. Like the wild bison, they remain together in a bunch, drop their nutrient-rich manure, push organic matter into the topsoil, and ready the land to receive and hold rainwater. Then the ranch manager herds them into the next paddock. The cattle never return to a grazing zone until the grasses have grown back.

Casey Wade, vice president of ranching operations for Dixon Water Foundation, maintains a detailed plan of how many animals will be grazing in which locations based on the amount of grass available and its rate of growth. The plan is never set in stone, however, because adaptation is a core principle of managed grazing. As rainfall, forage availability, and other conditions change, Wade adjusts the plan. It sounds laborious—and it is—but the goal is to boost profitability by growing more grass, rejuvenating lands previously overgrazed, and raising more

Casey Wade moves cattle according to his rotational grazing plan on the Dixon Water Foundation's Mimms Ranch outside of Marfa, Texas. Photo © Terrie Wade.

cattle per acre. "Bare ground is really the enemy," Potts said, echoing the words of North Dakota farmer Gabe Brown and USDA laboratory director Jerry Hatfield. "If you have one-third bare ground, your ranch is one-third smaller."

Currently, the Mimms Ranch supports 200 "animal units" (roughly equal to 133 cow–calf pairs) of purebred Hereford and Red Angus. Most of the cattle raised for sale graze for 24–28 months and are then sold to the Grassfed Livestock Alliance, which in turn supplies beef to Whole Foods Markets throughout Texas. On average each of Dixon's grass-fed steers brings in about twice what a conventionally raised steer would. In large part that's because health-conscious consumers prefer leaner meat to the fattier, marbled cuts from cattle fed corn in feedlots. Even the Cleveland-based Mayo Clinic notes that grass-fed beef contains nutrients and fats that can be beneficial to heart health and may reduce cancer risks.[21]

While cattle raised on irrigated pasture get the prized "grass-fed" label as well, those raised on natural rangeland under holistic management techniques offer a suite of other potential benefits. As we tour the Mimms Ranch, I learn that the land is alive with birds—quail, western meadowlarks, kestrels, vesper sparrows, and savannah sparrows, to name a few. The ranch's 270 different species of grasses and plants create a rich variety of habitats for birds and wildlife. As we bump along a rutted dirt road, Potts points to a chestnut-collared longspur, a small ground-feeding bird that is approaching threatened status internationally. Swainson's hawks nest on the ranch, and in his planning Wade will do his best to keep cattle out of those areas during nesting season.

For the nonprofit National Audubon Society, a conservation organization focused on bird conservation, ranches like Mimms that are managed for a synergy of ecosystem benefits and economic profits present a great opportunity. Native grasslands are among the most altered and imperiled ecosystems in the world—and one of the least protected. Throughout North America, grassland ecosystems are dwindling at a rapid rate due to poor grazing and agricultural practices, the proliferation of invasive plants, and encroaching human development. In the United States, populations of 24 grassland-dependent bird species have declined by nearly 40 percent over the last half century, according to Audubon. With 85 percent of grasslands in the United States privately owned, Audubon hopes to motivate more grassland conservation by offering a "bird-friendly beef" certification for ranches that meet its habitat criteria.[22]

"The idea is to incentivize restoration of grasslands at a landscape scale across Canada, the US, and Mexico," said Beth Bardwell, director of conservation for Audubon New Mexico, when we met over coffee shortly before my visit to the Dixon ranches. Our goal is "to create healthy grassland bird habitats through environmentally focused ranching protocols that benefit both ranchers and birds."[23]

Audubon hopes to pilot its "bird-friendly beef" certification on the family-owned Ranney Ranch, an 18,000-acre (7,300-hectare) spread outside of Corona, New Mexico. Once a classic example of overgrazed rangeland, the ranch has been transformed into a healthy grassland ecosystem by rotational grazing combined with the selective removal of opportunistic (and thirsty) juniper trees. Nancy Ranney, who studied landscape architecture at Harvard University, took over the family operation after her father's death in 2002 and had a strong desire to ranch holistically. When I visited the ranch in early July 2016, her 200 head were grazing in one of 35 paddocks that range in size from 20 to 2,000 acres (8 to 809 hectares). She sells her animals when they are 6 to 9 months old and weigh about 600 pounds (272 kilograms), roughly half what most cattle weigh when sold. "Our model truly fits the environment of the Southwest," Ranney wrote to me in a follow-up email after my visit. "It is the best use of our land and our resources, and it produces excellent beef."

The diversity of grasses on Ranney's rangeland has grown from 5 species in her father's day to 45 species today. As we toured the ranch, a mix of rolling hills and flat lands, patches of winter wheat waved in the breeze. "Once we started doing the rotational grazing, the cool-season grasses just started appearing," Ranney said. They make a big difference, she explained, since they extend the growing season by about two months. Rotational grazing has also boosted vegetative cover from about 40 to 70 percent, which means less erosion, more water infiltration, higher carbon content, and greater land productivity.

It's that healthier grassland landscape that makes for good bird habitat and that got Audubon talking to Ranney, who also serves as president of the nonprofit Southwest Grassfed Livestock Alliance, an organization that supports grass-fed livestock producers. A bird-friendly label might just give a marketing edge to grass-fed beef from cattle raised on natural rangeland as opposed to irrigated pasture. And it would teach consumers that well-managed ranchland can provide healthy habitats.

The creation of bird habitat is only one of many "ecosystem services" rotational grazing offers. Another is improved streamflow. During my visit to the Dixon ranches, Potts drove me south of Marfa to see Alamito Creek, which runs through a 7,500-acre (3,035-hectare) tract of rangeland also owned by the foundation. Alamito, which joins the Rio Grande just upstream from Big Bend National Park, is one of the healthier creeks in West Texas. The mixed-age stands of cottonwoods and willows along the creek's banks suggest that its flood regime is still at least partially intact. The threatened western yellow-billed cuckoo nests along the creek, and the spectacular vermilion flycatcher is so common that locals jokingly call it a "trash bird."

But previous overgrazing of the surrounding land has increased rainfall runoff, eroding and incising the creek's channel. This in turn has lowered the water table and dried out the surrounding rangeland. Creosote bushes and bare ground dominate, instead of grasses. While we saw an abundance of life in and around the creek—from fish and frogs to insects and turtles—Alamito was a series of pools rather than a flowing stream.

Potts believes managed grazing can change that. "It's all about holding more water in the soil," Potts said. When restored to health, the grassland will absorb rainwater and release it slowly to the creek. With controlled grazing, both the channel and the area around it will improve. Better land management, he believes, will help Alamito flow more continuously. "We're starting to get some recovery," Potts said, "but it's going to be a hundred-year process."

If the experience of Sid Goodloe on his Carrizo Valley Ranch in south-central New Mexico is any indication, it might not take quite that long. When Goodloe purchased his 3,450-acre (1,400-hectare) ranch in 1956, it was overrun with juniper, pinyon, and ponderosa pine. The land looked more like forest than the savannah grassland it once was. A creek bed that sliced through the property was completely dry. Goodloe was "just lucky" to meet Allan Savory, the founder of holistic management, while in Africa in the sixties. After spending time with Savory in the savannahs of Rhodesia, he brought what he learned home to New Mexico and began practicing rotational grazing on his ranch.[24]

Goodloe's land began to heal. Healthy grasslands returned. Creeks that once ran only after a decent rainfall began to flow year-round. The healthier watershed has attracted mule deer and elk, so in addition to selling ranch-raised, grass-finished Angus beef, Goodloe brings in revenue from hunting fees. He also sells lumber, firewood, and vigas. With this diverse income, he and his wife, Cheryl, can better weather drought, a fact of life in New Mexico. "We've found it profitable to work with nature," Goodloe says.[25]

Goodloe's pioneering work in New Mexico, along with Savory's cofounding of Holistic Management International in Albuquerque in 1984, helped to spread the gospel and practice of rotational grazing. But despite numerous examples of its successful use by ranchers in the United States and many other countries, the mainstream range-science community remains skeptical of its benefits. In part this is because a proper evaluation would require a geographic scale of whole landscapes and a time frame of many years, which most scientific grants do not support. In addition, rotational grazing's core principle of adaptive management—adjusting grazing plans as conditions change—makes it hard to study using conventional scientific methods. Or, as Potts put it, "It does not lend itself to a reductionist research approach."

The Dixon Water Foundation has supported the work of Richard Teague, a research ecologist at Texas A&M AgriLife Research in Vernon, Texas, who has conducted long-term trials on the foundation's ranches in North Texas. Teague's nine-year studies of grazing methods in Texas tallgrass prairie found that, compared with neighboring ranches where cattle grazed continuously (the conventional approach to western ranching), rotational grazing improved soil health, grass quality, nutrient availability, and water-holding capacity. His team also modeled effects on the watershed and found improvements in the quality and flow of local streams due to greater water infiltration and more than 30 percent reductions in the runoff of sediment, nitrogen, and phosphorus. With the land acting more like a sponge, there would be less risk of harmful flooding downstream.[26]

Teague's research also shows that, in contrast to the popular view of cattle as climate destroyers, they can actually help mitigate climate change. He and his colleagues found that on rangeland previously degraded by heavy continuous grazing, intensive rotational grazing increased the carbon content of the soil by an average of 3 metric tons per hectare (1.3 tons per acre) per year in the top 90 centimeters (3 feet) of soil over a decade. Managed in this way, cattle can help sequester enough carbon in the soil to more than offset their methane emissions.[27]

Despite its many benefits, rotational grazing won't be easy to scale up. The capital costs for water troughs and fencing can be substantial. At the Dixon ranch, installing an electric fence runs about $1 per foot, or $5,000 per mile ($3,100 per kilometer). Dixon ranch manager Jonathan Baize has invented an electric ear tag that he hopes will do away with the expense of fencing. If a cow wearing the device moves too far from the herd, she gets a warning tone; if she keeps moving away, she gets a shock. The idea is to quickly train cattle to stay bunched together. As of late 2016, the electric ear tag was under evaluation at the US patent office.

The second barrier is a cultural one. It takes time to learn how to create detailed plans, regularly move the animals, and monitor forage and land conditions. As ranch overseer Wade said, "You can't be a rowdy cowboy and make this work."

Among the biggest barriers, though, are federal, taxpayer-funded farm subsidies that incentivize landowners to plow up prairies and plant crops rather than bear the risks of raising cattle on the range. Between 2007 and 2012, the US area in rangeland dropped by about 1.5 million acres (607,000 hectares). While a number of factors drove that decline, government programs played a role. We're losing "huge amounts of grasslands due to these federal subsidies," said Potts. "If you're going to have farm programs, they've got to be weighted toward those that support good ecosystem practices."[28]

So, can a true environmentalist eat beef? As with so many questions of this sort, the answer starts with "It depends." But for those in search of an environmentally ethical meal, beef from range-raised and holistically managed cattle cuts the muster. That said, it's highly unlikely that ranchers could produce enough beef in this ecologically sound fashion to satisfy current demand. To be a sustainable part of our diets, beef might need to become, in the words of British farmer and author Simon Fairlie, "a benign extravagance."[29]

CHAPTER 7

Conserve in the City

Water, is taught by thirst.
—*Emily Dickinson*

BACK IN 2007, United Water decided to build a desalination plant to increase water supplies for customers in one of its service areas. The company, a US subsidiary of the multinational Suez Environment, predicted shortfalls as early as 2015 and proposed the desalting facility to fill the gap. On the face of it, there was nothing particularly unusual about the idea. Turning saltwater into drinking water has become big business in many parts of the world, from California to Australia, Saudi Arabia, and Israel. Today, more than 18,400 desalting plants dot the world's coastlines.[1]

But the customers in this case were some 300,000 residents of Rockland County, New York, about 30 miles (48 kilometers) north of New York City. Average annual rainfall there is a generous 49 inches (1,245 millimeters). United Water proposed pumping millions of gallons a day from Haverstraw Bay, a tidal area of the Hudson River, removing the salts and impurities, and then selling the water to Rockland's residents. The company estimated the plant's capital costs would run some $150 million, a cost that customers would bear through higher water rates.

Concerned that corporate profits were about to trump the public good, Rockland residents mobilized. A band of citizens and conservation organizations formed the Rockland Water Coalition, which included groups such as Scenic Hudson, Riverkeeper, and Food & Water Watch, to mount an offensive. With the support of local elected officials, the coalition spearheaded a campaign that highlighted desalination's high energy requirements and large capital costs, the proposed plant's impacts on the fisheries of Haverstraw Bay, which is among the most vital spawning and nursery grounds throughout the Atlantic Seaboard, as well as concerns that the plant's intakes would be located just 3.5 miles (5.6 kilometers) from discharges to the Hudson River by the Indian Point nuclear power plant. Most compelling, however, was the coalition's argument, bolstered by expert evidence, that the costly desalination plant was completely unnecessary.[2]

"We don't have a water shortage problem, we have a water management problem," said Laurie Seeman, the founding director of an arts-based environmental education organization and a leader of the Rockland Water Coalition. Seeman was particularly concerned about damage the proposed plant might cause to the fish and wildlife habitat of Haverstraw Bay, where she often brings students. "Not on my watch," she said.[3]

In July 2014, the *New York Times* reported that Albert F. Appleton, a former commissioner of environmental protection for New York City, had informed the New York Public Service Commission (PSC), the agency that would ultimately decide the desalination plant's fate, that even if projected demand for Rockland County was to exceed available supplies, "there are numerous alternative ways to meet it at a price that, at most, would be 10 percent of the cost of the desal facility." But more importantly, effective conservation might delay the need for new supplies for some time. As commissioner of the city's Department of Environmental Protection, the largest municipal water utility in the United States, Appleton had overseen a highly successful conservation program that continues to this day. Despite vibrant economic growth, the city's water use is now lower than it was a half century ago.[4]

In Rockland County, previous projections had suggested the need for new supplies, but those estimates overshot reality by a considerable margin. In 2006, the PSC predicted that by 2014 the county's water use would rise to 34 million gallons (129 million liters) per day, which turned out to be 20 percent higher than actual demand.[5]

That kind of miscalculation gets to the heart of water conservation's unsung success. In most cities and towns across the United States, household water use has gone down over the last two decades, in many cases enough to completely offset any rise in water use from population growth. The reason largely has to do with the national water efficiency standards that required plumbing manufacturers to reduce the volume of water used by toilets, urinals, faucets, and showerheads. The requirements became law as part of the 1992 US Energy Policy Act signed by US president George H. W. Bush. Since 1994, all new and remodeled homes and most commercial buildings and facilities have been required to install these efficient fixtures in all renovations and new construction, effectively building conservation in to urban and residential infrastructure. According to Amy Vickers, an engineer and nationally recognized water conservation expert based in Amherst, Massachusetts, who wrote the national efficiency standards, those standards are now saving an estimated 7 billion gallons (26.5 billion liters) per day—equivalent to seven times the daily water use of New York City.[6]

With the addition of efficiency standards for clothes washers and dishwashers, along with the 2006 launch of the US Environmental Protection Agency's water-efficiency labeling program called WaterSense, the savings only grew. According to a detailed analysis of more than 730 homes across the United States and Ontario, Canada, sponsored by the nonprofit Water Research Foundation in Denver, Colorado, indoor water use declined from 69.3 gallons (262 liters) per capita per day in 1999 to 58.6 gallons (222 liters) in 2016, a drop of 15 percent. In the homes surveyed, only 37 percent of toilets and 46 percent of clothes washers met the new efficiency levels in 2016, so water use will continue to fall as older homes are remodeled or replaced, and new fixtures and appliances get installed. In the coming decades, researchers expect indoor water use per person to drop at least an additional 37 percent.[7]

Proven methods also exist to reduce lawn and landscape irrigation and to fix leaks in utility distribution systems, as well as to turn local wastewater and storm water into new sources of supply. The failure to adequately account for the potential of conservation, reuse, and smarter management to cost-effectively reduce demand has often led water providers to propose unnecessary and expensive new supply projects, like dams and desalination plants. Private water companies that have shareholders to please often benefit most from these big projects because they can pass the capital costs on to their customers and then increase company revenues and profits by selling water at higher rates. For residents, however, such projects typically mean higher water bills and more harm to the environment. And this was precisely what the residents of Rockland County were determined to prevent.

"Everything was being called an alternative to desalination, when desalination should have been one of the alternatives," said Seeman of the Rockland Water Coalition. "What (United) termed education was a lavish advertising campaign for desalination. They attempted to power it through."

In November 2014, after years of steadfast organizing, outreach, and activism, the Rockland Water Coalition achieved a short-term win. State regulators ordered United to suspend plans for the desalination project and work with a new county water task force to evaluate conservation and other measures to close the gap between projected demand and supplies. For help analyzing its water use and management options, the county hired Amy Vickers, the conservation expert who had written the national efficiency standards and had also designed conservation plans for utilities around the country. Vickers's analysis showed that the county's water demand had largely been flat since 2000, despite an 11 percent increase in population. She estimated that tried-and-true residential conservation along with the repair of leaky water mains and pipes could reduce the county's water demand by 15–25 percent. Based on the success of programs in Boston, New York City, Seattle, and elsewhere, the savings could be even greater.[8]

One finding, in particular, spotlighted United Water's shortcomings: the large volume of water lost to leaks in the distribution system. In most years unaccounted-for water (also called nonrevenue water because it never reaches a billable customer) exceeded even the state's outdated and lax cap of 18 percent. (In most well-managed systems, water losses constitute less than 10 percent of demand.) United was a decade behind the state's recommended schedule for surveying water mains for leaks, and at its recent "sluggish pace" of replacing mains it would take up to 704 years to complete the job, according to the Vickers report. A local paper ran a cartoon depicting a United Water supervisor praising a team of futuristic alien-like workers: "Well, fellas, we're finally about to replace the last water pipe in Rockland and we did it a year ahead of schedule. It only took us 703 years."[9]

Upon release of Vickers's report, Rockland County legislator Harriet Cornell said the task force, which she led, could now move forward with efforts "to ensure a safe, cost-effective, long-term water supply for Rockland." On December 18, 2015, the PSC ordered United Water (which a month earlier had changed its name to Suez) to abandon the desalination plant and pursue alternatives. Against the odds, Rockland residents had prevailed.[10]

It turned out, however, that Rockland's David-versus-Goliath battle with United was not over. United (now Suez) initiated action to recoup $54.5 million in planning and engineering costs it claimed to have incurred for the now-canceled desalination plant. With interest over a 20-year payback period, the total sum flowing back to United Water from Rockland ratepayers would amount to some $110 million. The company had already sought and received PSC approval for rate hikes to cover $39.7 million in preconstruction costs for the phantom plant. At hearings before the PSC in June 2016, Rockland County legislator and water task force chair Harriet Cornell stated, "the ratepayers should not be forced to bear the burden of the business risk taken by United Water."[11]

In late June 2016, Rockland County filed suit in New York's Supreme Court in Albany against Suez New York (formerly United Water New York) and two state agencies, including the PSC. It alleges that Suez violated its statutory responsibilities for imposing unjust or unreasonable charges on ratepayers for its water service. "Rockland County residents should not be left holding the bag for an ill-advised and poorly managed project—especially one that no one wanted and one that was ultimately shown to be unnecessary," said Rockland County Executive Ed Day in a press release. "The Public Service Commission, which is supposed to be the watchdog for consumers, stood by and did nothing as Suez, our water company, spent $54 million on a plant that was never built," Day said. "In short, Rockland ratepayers are being hosed."[12]

"They spent our ratepayer money to battle us," said Seeman, with the Rockland Water Coalition. "They continue to do so to this day."

Some Rockland residents are now questioning whether a private corporation should manage public water, and are calling for a "Take Back Our Water" movement. "What happens with Rockland's water will set a precedent for the rest of New York, and could be significant nationally," the coalition stated in a news release.[13]

For Seeman, the desalination fight is about more than fair water rates, as important as those are. "How we respond at this time to water security shapes so many things in our lives," she said. "It's going to shape how we protect our streams, how we minimize energy consumption, where our storm water goes, how our buildings are constructed. We want to live in a place that shows signs of social and ecological intelligence—a place that is preparing for the next generation."

~

Cities around the world occupy about 2 percent of Earth's land surface, but they house more than half the world's population and generate a majority of its economic activity. To meet those highly concentrated demands, cities have historically reached farther and farther out from their urban cores to tap new water sources. Researchers estimate that the world's large cities, defined as those with at least 750,000 inhabitants, collectively transfer more than 500 billion liters (132 billion gallons) of water per day a total distance of some 27,000 kilometers (16,780 miles). Positioned end to end, the canals and pipelines transporting this water would stretch halfway around the world. The volume of water transferred annually is equivalent to the yearly flow of 10 Colorado Rivers.[14]

As a result, in addition to straining their own local supplies, many growing cities are stressing watersheds, rivers, and lakes far removed. According to a 2014 study, Los Angeles, California, ranks first in the world in cross-basin water transfers, importing some 8.9 billion liters (2.3 billion gallons) per day from the Colorado River and northern California rivers to satisfy the demands of its 13.2 million people. Boston (United States), Mumbai (India), Karachi (Pakistan), and Hong Kong round out the top five large cities that import the most water from watersheds other than their own.

Of large water importers, Boston, Massachusetts, which obtains most of its supply from the Quabbin Reservoir in the central part of the state, was among the first to discover the lesson Rockland County, New York, is now embracing: that conservation and efficiency improvements are the most cost-effective and least environmentally damaging ways of meeting new water demands. Back in the 1980s, as greater Boston's water use climbed toward the limits of its available supply, water planners began doing what most do: search for a new supply. The leading contender was a diversion from New England's largest river, the Connecticut, some 130 kilometers (80 miles) west of the city. But citizens concerned about the proposed diversion's impacts on the river and its fisheries banded together and pushed for conservation.

The upshot was a shift in focus from grabbing more water to using less. In 1988 Massachusetts amended its plumbing code to become the first state in the nation to require low-volume, 1.6-gallon-per-flush toilets. The Massachusetts Water Resources Authority fixed leaks in the aging distribution system, retrofitted 370,000 homes with efficient plumbing fixtures, conducted industrial water audits, improved water meters, and educated the public about conservation. Greater Boston's water use dropped 43 percent from its peak, and today is back where it was a half century ago. There is no longer talk of a long-distance river diversion.[15]

For Singapore, the Southeast Asian city-state that gained independence from the United Kingdom in 1965, dependence on water imports has created something of an existential threat. The island nation of 719 square kilometers (278 square miles) and 5.4 million people relies on water piped in from Malaysia for a portion of its drinking water supply. On occasion Malaysia has made veiled threats of withholding water to influence politics. One of two water-transfer agreements between the two nations expired in 2011; the second expires in 2061. To free itself from water dependence, Singapore has pursued a mix of strategies, including local rainwater capture, conservation and efficiency, wastewater reuse, and desalination.[16]

The PUB, Singapore's national water agency, follows a strategy of using the cheapest option first. With rainfall averaging a hefty 2,400 millimeters (94 inches) per year, storm-water capture constitutes an important part of the tiny nation's supply. Through a network of drains and canals, the water agency channels two-thirds of the island's urban runoff to 17 reservoirs, where it is then treated and added to the drinking supply. It has reduced leakage and unaccounted-for water to about 5 percent of demand, among the lowest rates in the world. Since 1994, conservation efforts have cut per capita domestic use by 14 percent, and the goal is to bring that usage down another 7 percent, to 140 liters (37 gallons) per person per day, by 2030.[17]

Along with conservation and leak repair, Singapore has pushed the envelope in the recycling and direct reuse of its wastewater. This so-called NEWater emerged on the Singapore scene around 2002 when a panel of experts approved the addition of highly treated wastewater to the public drinking water supply. A tunnel 48 kilometers (30 miles) long collects and transports sewage to reclamation facilities where it then receives treatment with advanced membrane technologies and ultraviolet disinfection, bringing it up to drinking quality. Strong public education and outreach fairly quickly overcame the "toilet to tap" concerns that cities elsewhere in the world have faced. Singapore's four NEWater plants can satisfy up to 30 percent of the nation's current demand. By 2060, a year before the expiration of Singapore's contract with Malaysia, reclaimed water is expected to meet up to 55 percent of that higher demand.

"Ideally, we don't sell you water. We rent you water," George Madhavan, communications director for Singapore's water agency, told *Reuters* in 2015. "We take it back, we clean it. We're like a laundry service. Then you can multiply your supply of water many, many times."[18]

For Singapore, desalination is a last resort because of its expense and energy requirements. But as a densely populated nation with a growing economy that is already efficiently tapping most of its natural supply and still depends on Malaysia for about half of its water needs, the city-state has had little choice but to turn to the sea. Two desalination plants meet 25 percent of current water demand, and with the addition of three more units in the coming years, the desalting of seawater is expected to meet 30 percent of demands in 2060. To keep costs and energy use down, Singapore hopes to site the last facility near an existing power plant.[19]

Cities in the much larger and drier island nation of Australia offer a cautionary note about turning to the sea too quickly. During the 14-year drought that ended in 2010 (see Chapter 10), Australian authorities came to view desalination as an important insurance policy against the next Big Dry. Nearly US$10 billion was invested in four big desalting facilities in Adelaide, Brisbane, Melbourne, and Sydney. But after the rains returned in 2010, the expensive desalted water wasn't needed. Conservation measures implemented before and during the drought not only helped cities weather the epic dry spell, they reduced water demands to levels that rendered the new desalination plants unnecessary. As of late 2016, none of the four plants were operating, yet their construction and maintenance costs must be paid. These stranded desalination assets, writes reporter Keith Schneider with the nonprofit Circle of Blue, "are a cautionary tale of growing too big too fast."[20]

The impulse to build a desalination plant that can be flipped on during a prolonged dry spell is understandable. After all, it's the utility's job to supply water to customers during droughts, floods, and every condition in between. It's also their job, however, to supply that water without driving up costs or harming the environment. Fossil-fueled desalination plants contribute to climate change, which increases the risk of drought—a less-than-elegant solution, to say the least.

Australia's experience offers a clear lesson for forward-thinking utilities: conservation is a less expensive and more sustainable solution than elaborate engineering projects. While some emergency drought measures, such as outdoor watering restrictions, are temporary, most conservation measures reduce urban and residential water use permanently—and they save money. Stuart White, director of the Institute for Sustainable Futures at the University of Technology Sydney, and his colleagues have estimated that the unit costs of conservation and efficiency programs are typically 27–38 percent that of desalination plants. Even wastewater reuse, one of the more expensive options, costs about 30 percent less than desalination.[21]

On the supply side of the equation, cities can often lease or buy agricultural water rights, or invest in on-farm efficiency in exchange for the water saved, for a fraction of the cost of desalination or new dam construction. Managers can still prepare plans for engineering projects should they become needed—but the principle of starting with the lowest-cost strategy still holds.[22]

In addition to conservation, cities around the globe are beginning to mine the same liquid gold Singapore has tapped for decades: urban storm water. Cities that were long plagued with flooded streets, eroded streams, overflowing sewage systems, and polluted rivers and bays are transforming urban runoff from a nuisance to an asset. Portland, Oregon, for example, has invested in "green roofs" to capture rainfall and "green streets" to soak up runoff in order to prevent sewer overflows into the Willamette River. Chicago, Illinois, now boasts more than 500 green roofs—including one atop City Hall—that collectively cover some 7 million square feet (650,000 square meters). The vegetated roofs catch rainwater, beautify and cool the city, and expand space for urban gardens.[23]

It's all about designing in nature's image and allowing rain to do what it did before buildings, roads, driveways, parking lots, and other impervious surfaces covered the landscape: soak into the earth, fill the soil, recharge groundwater, and release flow gradually to rivers and streams. Along the way natural systems filter and cleanse the water. A large-scale pivot to green infrastructure requires a union of hydrology, ecology, and climate science with civil and environmental engineering. If done right, this approach will help cities weather the coming age of floods, droughts, and water stress.

Los Angeles is taking a page from nature's playbook and making better storm-water management, along with conservation, a cornerstone in its efforts to reduce long-distance water imports and, by 2035, to source half of its supply locally. The City of Angels has committed to capturing more than 61,300 acre-feet (20 billion gallons or 76 billion liters) of storm water per year by 2025. Richard Luthy, a professor of civil and environmental engineering at Stanford University, estimates that with sizeable community-scale projects Los Angeles could capture as much as 100,000–200,000 acre-feet (123–246 billion liters) of storm water per year. "Those are big numbers," Luthy said. With LA's water needs projected to total some 700,000 acre-feet (863 billion liters) by 2035, the capture and underground storage of storm water could potentially meet 14–28 percent of the city's water needs. And that means avoiding the need for expensive and ecologically harmful desalination.[24]

Cities in the eastern half of the United States are also turning to green infrastructure for storm-water management. Leaving aside the benefits of capturing and saving rainwater, no one wants untreated sewage flowing into local rivers or bays. In 2016, the Institute for Sustainable Infrastructure recognized KC Water in Kansas City, Missouri, for constructing wetlands, rain gardens, pervious pavement, and other green infrastructure projects to prevent sewer overflows, which also improves water quality in the Middle Blue River.[25] Philadelphia, Pennsylvania, plans to invest some $2 billion in the coming years in what is hailed as the most comprehensive green infrastructure plan for storm water in the United States. The city has designed or implemented nearly 1,000 projects that encourage rainfall to infiltrate rather than run off the cityscape, including tree trenches, rain gardens, capture basins, vegetated swales, wetlands, and the installation of pervious pavement. Philadelphia's goal is to reduce storm-related sewer overflows by 85 percent over 25 years.[26]

For Copenhagen, Denmark, a low-lying northern European nation situated between the Baltic and North Seas, greening its cityscape is closely tied to preparing for climate change. The Danish capital experienced two 100-year floods in the last six years, more than hinting at the "new normal" that lies ahead. Instead of upgrading its drainage pipes and other "gray" infrastructure, Copenhagen is expanding and redesigning parks and other public spaces to capture and store more storm water. The urban park called Enghaveparken will channel runoff to 100 community gardens, and during big storms some of its athletic fields will double as catchment basins. Overall, the city's $1.3 billion investment in green infrastructure is half the price of a more conventional gray-infrastructure approach. Plus, it improves Copenhagen's quality of life.[27]

For its part, China has adopted a creative term, *sponge cities*, for smarter urban water management. With 19 percent of the world's population but only 7 percent of its renewable freshwater, China faces water shortages on a vast scale. But floods are also a growing challenge. By late July, the year 2016 had already surpassed 1998 for the costliest year of flooding on record, with damages totaling nearly $45 billion, according to *China Daily*. The extent and pace of China's urbanization are unmatched in world history. Over the past 35 years, the number of cities in China has climbed from 193 to 653, and the urban landscape has grown by some 40,000 square kilometers (15,400 square miles)—an area 55 times that of Singapore. Vast areas of lakes, wetlands, and woodlands have been replaced with buildings, roads, and parking lots. Not surprisingly, drainage systems have not kept pace with the swells of runoff. In 2013, severe flooding hit some 230 Chinese cities.[28]

That was the year Chinese President Xi Jinping proclaimed at a conference on urbanization that cities should act like sponges, absorbing rainwater instead of allowing it to surge down streets and sidewalks. In 2015, the government selected pilot sites in 16 cities—including Beijing, Chengdu, Guangzhou, Shanghai, Shenzhen, and Wuhan—to test the idea. Each sponge city receives 1.2 billion yuan ($175 million) from the central government over the first three years and is expected to supplement this funding with local and private investments to undertake projects that absorb rainfall, capture storm water, and recharge aquifers and other local water sources. In a newly urbanizing area of Chongqing, for example, urban designers are installing permeable pavement, allowing rainwater to soak into the earth below. Plants and trees rather than painted lines define parking spaces.[29]

It's too soon to tell whether China's sponge city idea will catch on and expand. Compared with Copenhagen or Philadelphia, the investment per city so far is modest. More funding will be required to meet the Chinese government's goal that one-fifth of each pilot city meet sponge-city standards by 2020. But the concept is forcing Chinese planners to rethink how water moves through their urban landscapes, and with floods and droughts likely to worsen, that rethink is coming none too soon.[30]

～

It was a sparkling, blue-sky mid-November day, with the golden hue of autumn cottonwood leaves outlining the banks of the Rio Grande in New Mexico, as I ventured out with Katherine Yuhas, manager of the water resources division for the Albuquerque–Bernalillo County Water Utility Authority. Within a half hour of leaving Yuhas's downtown office in City Hall, we arrived at an arroyo called Bear Canyon in the city's residential northeast quadrant. Just to the east were the foothills of the imposing Sandia Mountains and to the west the beautiful Rio Grande valley. Under natural conditions, the arroyo would be dry at this time of year, but on this day it was a shallow creek flowing slowly past a public playground and golf course. It sparkled as it cascaded downhill in step-like fashion for nearly half a mile.

"The community loves it," said Yuhas, who came to the water authority in 2003 after serving as hydrologist for Santa Fe County and groundwater protection specialist with New Mexico's environment department. But the reason Yuhas brought me to Bear Canyon is that it's much more than an amenity: it's a cornerstone of Albuquerque's ambitious plan to secure enough water to meet the city's needs for 100 years.[31]

A century is a long time. Looking back that far, Model T Fords had just begun to nudge out horse-and-buggies, and the Wright brothers' first flight in a powered aircraft was still a fresh memory. Certainly no one back then could have guessed so many of us would be jet-setting around the world and sending videos instantaneously across the globe through something called cyberspace. It's anyone's guess what the world of our descendants will look like 100 years from now. But one thing is certain: every inhabitant will need water.

That fact is the motivation behind Yuhas's work at the Albuquerque water authority. In a region where precipitation averages a meager 9.4 inches (239 millimeters) per year, and with New Mexico already battling Texas in court over their respective rights to the Rio Grande, securing enough water for Albuquerque residents in the year 2120 sounds like mission impossible. But Yuhas says it's not only possible, it's done: "We don't need any new water for 100 years. We don't plan to buy any more rights or import any more water."

For a city that was told in the early 1990s that it was at serious risk of running out of water, that's a startling statement. Albuquerque's wake-up call came in 1992 when scientists with the US Geological Survey discovered that its aquifer, the sole source of drinking water for the city's then 400,000 residents, held substantially less water than previously believed. Follow-on studies showed that Albuquerque was pumping its groundwater twice as fast as nature could replenish it. The city was not only shrinking the flows of the Rio Grande, it was depleting its aquifers by 67,800 acre-feet (83 billion liters) per year. (At today's usage levels, 1 acre-foot can supply about three homes in Albuquerque for a year.) This news was not good. It conjured images of abandoned neighborhoods and shuttered stores.[32]

Rather than sink into denial, Albuquerque mobilized. The first order of business was putting in place an aggressive conservation program to reduce water demand, efforts that continue to this day. The utility fixed leaks in its distribution system and has brought losses down to an impressive 7 percent of system-wide demand. It offers rebates to customers to switch out older fixtures like water-guzzling toilets and clothes washers for more efficient models. It also educates residents about outdoor watering. Previously, Yuhas said, many customers tended to irrigate "as if it's July all year long." In her earlier role as the water authority's conservation manager, she designed a "Water by the Numbers" program that ingrained the irrigation code 12321 in everyone's mind. It instructs residents to water no more than once per week in March (the beginning of the irrigation season), twice per week in April and May, three times per week in June, July, and August, and so on. The catchy algorithm works. "Everyone knows it," Yuhas said.

Reducing landscape irrigation is particularly important because outdoor use is "consumptive," meaning that water evaporates or transpires through plants back to the atmosphere and is therefore unavailable to use again or to return to the Rio Grande. The Albuquerque water authority also pays $1.00–$1.50 for every square foot of green lawn converted to drought-tolerant landscaping, often called xeriscape (from the Greek *xerós*, meaning dry). To date the rebates have motivated residents to replace more than 2 million square feet (185,800 square meters) of turf with beargrass, winter jasmine, and other water-thrifty plants. Similar "cash for grass" programs have reshaped landscaping in many other western US cities from Austin and San Antonio, Texas, to

Denver, Colorado, and San Diego, California. The rebate program in Las Vegas, Nevada, has helped convert more than 179 million square feet (16.6 million square meters) of lawn to low-water-use landscapes. With every converted square foot estimated to save an average of 55 gallons per year, the Las Vegas program annually recoups some 9.8 billion gallons (37.1 billion liters).[33]

Like other forward-looking utilities, the Albuquerque water authority prices its water according to an increasing block rate structure. Because the unit cost of water rises along with water use, customers have an incentive to conserve. Of course, to pay for new infrastructure and maintenance, water rates overall must occasionally be increased. While rate hikes are never popular, customers who respond to them by using even less water might see no increase in their water bills. Others, however, may see an increase, and if any complain Yuhas has a ready response: "How much is your cell phone bill? How much is your water bill? If I told you that you were going to lose one (service) tomorrow, which one would you want to do without?"

Two decades of steady and strategic conservation in Albuquerque has paid off in spades. The city's total water demand per capita (including residential, multifamily, commercial, etc.) has dropped 49 percent since the conservation program began in 1995. Today, the city's residential water use stands at 92 gallons (348 liters) per capita per day, low for a semiarid city, and it continues to fall. (The average for the nation as a whole is 88 gallons [333 liters] per capita per day, according to the US Geological Survey.) Moreover, the share of residential water used outdoors has fallen from 60 percent to 40 percent. Conservation effectively halted depletion of the city's aquifer by 2000 and then held water levels steady for another eight years.

Groundwater levels actually began rising in 2008, when Albuquerque completed the infrastructure needed to tap a new supply: water imported from the Colorado River basin. New Mexico is one of seven

US states with territory in the Colorado watershed, and as a signatory to the 1922 Colorado River Compact it is entitled to some of the basin's water. Each year, a diversion from the San Juan River on the other side of the Continental Divide in the Colorado basin moves water into the Chama River in the Rio Grande basin for use in New Mexico. Thanks to this interbasin diversion, Albuquerque, which receives its San Juan–Chama deliveries via the Rio Grande, now gets about 60 percent of its supply from surface water. The reduced groundwater use has allowed aquifer levels to rebound, and Yuhas expects levels to continue to rise for another 20 years. The city is also storing some of its San Juan–Chama water underground, which gets us back to the sparkling arroyo at Bear Canyon.

Partway down the earth-and-rock embankment, water gushes out of a pipe into the creek below. Nearby sits a big beige tank marked with a band of purple, the color used by utilities to designate water as nonpotable (not for drinking). The tank holds surface water piped in from the Rio Grande, which now includes San Juan–Chama water. From mid-October to March, the water authority releases about 3 million gallons (11.4 million liters) a day from the tank into the arroyo. The water takes about 50 days to infiltrate the 500 feet (152 meters) down to the aquifer, where it will remain until needed. This "aquifer storage and recovery" project is an important component of Albuquerque's 100-year plan.

As we head away from Bear Canyon, the pieces of Albuquerque's water puzzle start to come together. Doing what the city is doing now, and assuming a medium degree of impact from climate change, the city would have enough water to meet its needs until 2085, says Yuhas. But the goal is adequate supply for a century. More conservation will close part of that gap. Reuse of wastewater, which gets treated at a facility south of the city, our next stop, will put Albuquerque over the finish line.

We're greeted by the water authority's chief engineer, Jeffrey Romanowski, who hands us hard hats and then escorts us to a large room filled

The water authority for Albuquerque, New Mexico, releases water from the Rio Grande into the Bear Canyon arroyo, where it replenishes the aquifer below. Photo courtesy of Albuquerque–Bernalillo County Water Utility Authority.

with big purple pipes and a series of deep rectangular concrete basins. Here wastewater that has already undergone conventional biological treatment gets pushed through fibrous pile-cloth media filters for an extra round of pollution removal and "polishing." The advanced filtration unit, which came online in 2012, brings the wastewater's quality to a level suitable for offsite reuse. When fully built, it will have the capacity to treat 29 million gallons (110 million liters) of Albuquerque's wastewater per day, nearly two-thirds of the city's current wastewater volume. (Albuquerque's current permit allows for the reuse of one-quarter of this capacity.) The water authority plans to build an 8.4-mile (13.5-kilometer) pipeline to connect the filtration plant with Bear Canyon so as to irrigate more parks and recreational areas with reclaimed wastewater and dedicate a greater portion of the San Juan–Chama water to recharging the aquifer.

In about a decade, the city plans to begin an "indirect potable reuse" project whereby some of its treated wastewater will be stored above- or belowground and then added to the city's drinking water supply. "We're going to use more (treated wastewater) for irrigation, put more in storage, and in 15 or 20 years we'll be drinking it," Yuhas said.

Rounding out Albuquerque's 100-year plan is protection of the forested watersheds where much of its surface water supply originates. After the impacts of the Las Conchas megafire in 2011, the authority barely had to think twice about signing on to the Rio Grande Water Fund, a collaborative effort to rehabilitate the region's watershed (see Chapter 3). With more big burns likely in the decades ahead, watershed protection is a critical leg of the water supply stool. "In the same way we do maintenance on our constructed reservoirs," Yuhas said, "we should do maintenance on our forests to protect those natural reservoirs."

In the late spring of 2016, armed with a slide show of charts and graphs depicting the authority's long-term strategy of conservation, wastewater reuse, aquifer storage, and watershed restoration, Yuhas and her colleagues took the 100-year plan to the public. They held a series of "customer conversations" throughout the city. The room was packed each time. Over light suppers of sandwiches and chips, the community heard the water authority explain the proposed plan and had a chance to ask questions and offer comments. Finally, at a town hall meeting on July 22, 2016, the public gave *Water 2120* a thumbs-up; two months later the authority's board officially adopted it.

To be sure, Albuquerque's population and economy have not grown at the breakneck pace experienced by Denver, Las Vegas, or Phoenix. Those cities have variously implemented some combination of conservation, reuse, and aquifer recharge, but continue to look to imports from distant water sources, the purchase of agricultural water rights, or both, to meet future needs. The fact that Albuquerque has no need to buy water rights from farmers in the Rio Grande valley has helped neighboring rural communities feel comfortable with the 100-year strategy. It also means that future water savings in agriculture could help restore needed flow to the Rio Grande.

It's a promising picture of water security. Especially in light of climate change, the plan is "prudent and gives our customers confidence," Yuhas said. "It's about providing flexibility for future generations."

CHAPTER 8

Clean It Up

The health of our waters is the principal measure of how we live
on the land.

—*Luna Leopold*

IT WAS A SWEET LITTLE JACK RUSSELL TERRIER named Rosie who
unwittingly became the proverbial canary in the coal mine for con-
taminated water in Long Island, New York. Annie and John Hall had
just finished lunch on their deck on a sunny day in early September
2012 when they found their beloved pet in a state of toxic shock near
the shores of Georgica Pond, where they'd lived and spent happy summers with their children and grandchildren for three decades. The Halls
rushed Rosie to the vet, but she was able to hold on for only a few days.
Scientists analyzed her tissue and found the culprit: microcystin, a toxin
produced by blue-green algae that is so potent the US military lists it as
a potential agent of biological warfare.[1]

Georgica Pond is nestled between the towns of East Hampton and
Wainscott in Suffolk County on Long Island's south shore. Spanning

some 290 acres (117 hectares), the pond is separated from the Atlantic by a sandbar only about 100 feet (30 meters) wide, its freshwater mixing with the salty sea. Over the decades, as the Hamptons became a prized summer retreat, large estates sprang up around the pond. Today, 74 homes ring Georgica's shores, including properties owned by filmmaker Steven Spielberg and billionaire businessman Ronald Perelman. Altogether some 4,000 acres (1,619 hectares) dotted with 2,000 dwellings drain into the pond.[2]

And therein lies the problem. Like two-thirds of Suffolk County's 1.5 million residents, those living near Georgica Pond use conventional septic systems that are incapable of thoroughly cleaning water from toilets. Human waste is chock-full of nitrogen, an essential nutrient, which, in excess, causes serious pollution and health problems. Nitrogen from both underground septic systems and lawn fertilizers migrates from backyard soils into Long Island's local streams and aquifers, which then flow into coastal lagoons and bays. There the nitrogen feeds algae, turning clear water into a green, fetid mess. Some algae release dangerous toxins, such as the microcystin that killed Rosie, but even nontoxic algae cause trouble when they decompose, robbing the water of oxygen and effectively suffocating fish and other aquatic life. This state of low oxygen, or hypoxia, can produce what are appropriately called dead zones.

Over the last several decades, algal blooms have become a growing threat to Long Island's ecosystems and economy. Eelgrass habitat, which is critical for scallops, oysters, and clams, has declined by nearly 90 percent as algae block out needed sunlight and housing and commercial development eat away at the coast. Fish kills are on the rise as algae infestations deplete oxygen levels in bays and estuaries. In late May 2015, hundreds of thousands of silvery-blue Atlantic menhaden, a type of herring, were asphyxiated when spiking water temperature, a large algal bloom, and an unusual concentration of fish conspired to suck oxygen out of the waters of eastern Long Island's Peconic Estuary.

Dead fish piled up in great masses along the shore. That same year, the presence of toxic microcystin forced the closure of 14 lakes and ponds in Suffolk County, including Lake Ronkonkoma, the island's largest.[3]

Algae are hurting not only fish but also people. A species called *Alexandrium,* creator of the notorious "red tides," produces a toxin that damages or even destroys nerve tissue. Scallops, mussels, clams, and other shellfish that feed by filtering particulates out of the water inadvertently draw in the toxic algae and concentrate them in their tissue. Eating these toxic shellfish can cause paralytic shellfish poisoning (PSP) with symptoms of numbness, dizziness, nausea, and loss of coordination. In severe cases, victims experience respiratory failure and die within a day. While species of *Alexandrium* have been present in Long Island waters since the 1970s, the first harmful bloom occurred in 2006. In the years since, the state has regularly closed shellfish beds around Long Island's coastal waters due to the risk of PSP, including as many as 13,000 acres (5,260 hectares) in a single summer. Researchers have determined that nitrogen flowing into coastal waters makes these red tides more dangerous.[4]

In Long Island and other coastal communities where fishing, recreation, and tourism support the local economy, algal blooms are as bad for business as they are for health. Preliminary assessments by researchers at Stony Brook University suggest that if nothing is done to reduce nitrogen pollution, the costs to fishing, tourism, and real estate on Long Island could total some $25 billion over the next 30 years.[5]

Long Island is by no means alone. Lakes, bays, and estuaries around the world are being overloaded with nitrogen and phosphorus, a process known as eutrophication. Whereas on Long Island the primary problem is household septic systems, in most regions, it's fertilizer runoff from farmland. Other sources include lawn and golf course fertilizers, animal waste from concentrated livestock operations, discharges from centralized treatment plants, and pollution from fossil-fueled power plants.

Globally scientists have identified more than 400 dead zones, and the number has risen rapidly in recent decades. Some of the most productive coastal fisheries are being threatened by toxic algae and oxygen-starved waters. Fish have died en masse, including in the German Bight of the North Sea, the Kattegat Sea between Sweden and Denmark, the northwest shelf of the Black Sea, Mobile Bay off the coast of Alabama, and Pamlico Sound off the coast of North Carolina, as well as the Peconic Estuary on eastern Long Island.[6]

Dead zones can be vast, often exceeding 5,000 square miles (13,000 square kilometers), and regularly form in the Gulf of Mexico off the southern US coastline, the East China Sea, and the Baltic Sea in Europe. Large lakes are being affected as well, with algal blooms springing up in the Great Lakes, the Great Salt Lake of Utah, Canada's Lake Winnipeg, and numerous lakes across Africa, China, and India.

In late June 2016, Governor Rick Scott of Florida declared a state of emergency in counties along the Atlantic Coast as an "unprecedented" outbreak of blue-green algae swamped long stretches of beach with foul-smelling toxins that some residents blamed for skin rashes. The algal bloom occurred after water was released to avoid flooding from Lake Okeechobee in the center of the state. The water from Lake O, as Floridians call it, was shunted toward the St. Lucie Estuary, one of the most biologically rich estuaries in North America. The previous month, NASA satellite images showed an algal bloom in Lake O covering some 33 square miles (85 square kilometers). Tests conducted later that summer by officials in Martin County, one of the coastal counties under a state of emergency, found levels of microcystin around the St. Lucie Estuary that were 100–1,000 times higher than the recreational guidance level. The toxin even showed up in air samples.[7]

"The alarm bells have rung. People want solutions," said Eric Eikenberg, CEO of the Everglades Foundation in Palmetto Bay, Florida.[8]

Alarm bells are also sounding in Toledo, Ohio. In early August 2014, officials warned the city's half million residents not to use their tap water for drinking, cooking, brushing teeth, or filling their pets' water bowls. A toxic algal bloom in Lake Erie had settled over the city's water intake pipe, and microcystin had been detected in the drinking water. Written off as ecologically dead in the mid-twentieth century, Lake Erie has made a comeback thanks to passage of the 1972 federal Clean Water Act, which curtailed the discharge of untreated industrial wastewater and other "point" sources of pollution. But the act did not control pollution from farms, lawns, septic systems, and other "nonpoint" sources, which allowed nitrates and phosphates to continue to flow into the nation's waters.[9]

Today more than half the world's people live within 60 miles (100 kilometers) of a coastline, and this share is expected to rise over the coming decades. Worldwide use of nitrogen fertilizer has climbed more than eightfold over the last half century, and continues to rise, particularly in developing countries.[10] And because blue-green algae like warmer water, global warming is expected to make the problem more serious and pervasive.

"Climate change will make the blooms worse," said Timothy Davis, a molecular ecologist with the National Oceanic and Atmospheric Administration's (NOAA's) Great Lakes Environmental Research Laboratory in Ann Arbor, Michigan. Microcystin, the toxin affecting Lakes Erie and Okeechobee and that killed Rosie along Georgica Pond, also does well in warmer water.[11]

With higher temperatures and warming water bodies, we're creating "a perfect incubator for these organisms," Davis said. "These events have been increasing all over the globe. They are becoming much more common." The situation, he said, "is alarming."

While the problem is global, solutions must be tailored to each locale. "The blooms are a visual representation of the problem of nutrient overenrichment," Davis said. "The most sure-fire way of making sure a bloom does not occur is making sure you have a balance of nitrogen and phosphorus in your system."

Achieving that balance will not be easy. It will require farmers, home-owners, landscape managers, and virtually anyone dealing with poop, pee, or fertilizer to do their part to keep nitrogen and phosphorus from entering rivers, lakes, and groundwater. Governments must mandate or provide incentives for landowners to adopt practices and technologies that reduce nutrients. It's a big challenge. But on Long Island, in the vast Mississippi River basin that drains into the Gulf of Mexico, and in other parts of the world awakening to the growing threat of algal blooms, solutions are coming to the fore.

~

On a beautiful Sunday morning in late August 2016, a flotilla of cars passed through the gates of the 57-acre (23-hectare) estate of busi-nessman Ronald Perelman and psychiatrist Anna Chapman on Long Island's East End. Four years after Rosie the terrier died from ingesting microcystin, dozens of people were gathering to hear the latest update on the quality of Georgica Pond. The Friends of Georgica Pond Foun-dation, a nonprofit formed in 2015, had hired Stony Brook University marine biologist Christopher Gobler to conduct a two-year study of the pond's ecosystem and what could be done to improve its health. It was time for Gobler to reveal his second round of findings. In attendance were Chapman, a director of the foundation, as well as local politicians, community members, and conservationists. Annie and John Hall were seated directly in front of me.

Gobler, whose trim frame and youthful face belie his 45 years, is a leading researcher on algal blooms and Long Island's water quality. He runs a research lab at Stony Brook's Southampton campus, where he investigates toxic algal blooms, fish die-offs, and the effects of climate change on marine ecosystems. A native Long Islander, Gobler recalls a "life-changing experience" a couple decades ago when a third-generation fisherman told him that the clams were disappearing in the Great South Bay, premier fishing grounds off Long Island's south shore. Since then, Gobler has been committed to improving the island's coastal waters, including, thanks to funding from the foundation, Georgica Pond.

For the previous three consecutive summers, the pond, one of Long Island's most productive breeding grounds for blue-claw crabs, was closed to fishing, crabbing, and swimming. As Gobler's investigations were getting under way in the summer of 2015, thick mats of floating, brownish macro-algae blanketed the lagoon. Nighttime oxygen levels were dropping to zero, and killing fish. (Algae produce oxygen during the day through photosynthesis and consume it at night through respiration.) Gobler's team installed a telemetry buoy on the pond to monitor its water quality in real time. By the end of August 2015, there were numerous recordings of extremely low oxygen. The team's early assessment indicated that more than half of the nitrogen fueling the algal growth was coming from local septic systems.

In 2016, Gobler reported, the pond was in much better shape than the previous summer. Early in the year, excavators had cut an opening in the sand barrier between Georgica Pond and the Atlantic, which allowed seawater to flush through the lagoon. Blue-green algae prefer salt levels less than 10 kilograms per gram, so as the water gets saltier, the algae population declines. "Opening the cut is like pulling the drain," Gobler said. Much of the nasty algae dissipates or moves out to sea. This temporary solution is not always possible, however, because the endangered piping plover nests on the sand barrier, and its habitat cannot legally be disturbed. But with the cut open for the first three months of 2016, the pond got a helpful cleansing.

Then in July the Georgica Pond team set about removing some of the vast mats of macro-algae. A mobile harvester dubbed the SS Georgica roamed the pond and scooped up 11 tons of algae, which, according to Gobler's calculations, removed 25 percent of the phosphorus and 10 percent of the nitrogen from the pond ecosystem. The algae would be tested to determine if it might be turned into a marketable fertilizer, but at a minimum, its removal improved the pond's water quality. While state environmental regulators had closed eight ponds on Long Island's south fork that summer due to dangerous algal blooms, Georgica was not among them. The algal harvesting "has been a huge success," Gobler said.

While helpful, flushing the pond with seawater and harvesting big mats of algae are not enduring solutions. In the near term, the Friends of Georgica Pond Foundation plans to invest $100,000 toward the installation of a "permeable reactive barrier," made of screens filled with wood chips, that is designed to trap nitrogen in the groundwater before it enters the pond. (The Massachusetts town of Orleans on Cape Cod is experimenting with this technology, as well.) The barrier may capture some of the nitrogen coming from high-priority areas close to the pond, but it won't stop widespread pollution from septic systems throughout the basin. And that is what must ultimately happen to protect Georgica Pond.

"Everyone is watching," said Nancy Kelley, the Long Island director of The Nature Conservancy, a partner in the foundation's work, at the conclusion of the Sunday morning session. The challenge at Georgica Pond is one much of Suffolk County faces, she said. "What we do here will have great impact on how we address this problem."

∽

As a little kid growing up on Long Island all I knew about our drinking water was that it came from the tall tower that loomed above our village of Franklin Square. I had no idea that the source of water for my backyard wading pool, Saturday night baths, and summer lemonade was beneath my feet. But it was, and that made our water vulnerable to so many things we Long Islanders did on the land above it—from applying fertilizers to our lawns and dumping waste in landfills to paving over wetlands for shopping malls and failing to properly treat and dispose of our sewage.

Over time, as demand for water climbed, the island was hit with another problem that plagues many coastal areas: saltwater intrusion into freshwater aquifers. As water is pumped from underground, the hydraulic gradient can shift: instead of groundwater flowing to the ocean, seawater invades inland aquifers. In Long Island, saltwater is contaminating the sole source of drinking water. A 2016 study in the journal *Science* by researchers at Ohio State University and NASA's Jet Propulsion Laboratory determined that 9 percent of the continental US coastline is at risk of saltwater intrusion, and 12 percent of the coastline is at risk of ocean contamination from septic tanks and fertilizer runoff; Long Island is at risk of both.[12]

Today, nearly 3 million people call Long Island home, roughly half in each of the two counties that make it up—Nassau, closest to New York City, and Suffolk. Shaped like a fish (or an alligator, depending on your point of view), the island stretches 118 miles (190 kilometers) long from New York Harbor east to Montauk Point and 23 miles (37 kilometers) wide from Long Island Sound on the north to the Atlantic Ocean on the south. Some 80 percent of Nassau County, where I grew up, is paved or in some way developed; most of the wetlands and open space are gone.

My childhood neighborhood, like many in the area, consisted of nearly identical houses on small lots—far too small for the installation of on-site septic systems. The household wastewater from most neighborhoods on the western end of the island was collected through sewer pipes and delivered to a centralized treatment plant. But in Suffolk

County to the east, properties tended to be larger and more spread out. Development came slower and considerably later, and even today much of the county remains quite rural, with vineyards and vegetable farms scattered across the landscape. Installing and operating centralized sewer systems gets expensive when homes are far apart, so most properties in Suffolk were equipped with septic systems that treat household sewage right on-site.

Septic systems are installed underground and typically consist of a tank and a drain field. Pipes deliver toilet waste and drain water from sinks and showers to the tank. Solids settle to the bottom while the liquid flows out of the tank and into the perforated piping that makes up the drain field. Naturally occurring bacteria break down the suspended solids as the water filters through the drain field toward the groundwater below. While basic septic systems provide a first line of defense against the spread of disease associated with human waste, they were not designed to remove nitrogen from the waste stream.

Today about one-fifth of US households have septic systems. As a simple, low-cost way to deal with sewage, they can do a decent job at protecting public health and the environment—if they're well maintained, but that's a big if. When they're neglected or situated too close together, and especially if groundwater is relatively shallow, as it is on Long Island, nitrates and other pollutants can contaminate drinking water and flow into lakes and coastal waters. That's precisely the problem now plaguing Suffolk County, where more than 7 in 10 homes use septic systems—the highest concentration in the United States.

The federal Environmental Protection Agency sets a limit of 10 milligrams of nitrate and 1 milligram of nitrite per liter in drinking water to guard against health hazards, including blood disorders in infants—the so-called blue baby syndrome. Nitrogen levels in Long Island's aquifers have shot up in recent decades, raising concerns that if nothing is done the island's groundwater—its sole source of drinking water—will begin to exceed the cap. Meanwhile algal blooms will worsen and spread.

"This has been a sleeper problem for a long time," said Jennifer Garvey, associate director of the Center for Clean Water Technology at Stony Brook University. "Many people don't even know what a septic system is."[13]

For most of Suffolk County, centralized sewer systems are not a realistic option. The county received some $383 million in federal funds to build central sewer connections for about 10,000 homes damaged by Hurricane Sandy—$38,300 per home. At that rate, building sewer infrastructure for the estimated 209,000 homes with septic systems situated in priority areas for nitrogen reduction would cost $8 billion.[14]

But the county is actively pursuing other options. "The magnitude of the problem created an opportunity," Garvey said, and that's to "accelerate innovation" in on-site wastewater treatment.

The center that Garvey helps direct was formed in October 2014 to do just that. With research and development support from state, federal, and private sources, the center aims to develop nonproprietary, advanced treatment technologies that can achieve a three-part goal: treated effluent that contains no more than 10 milligrams of total nitrogen per liter, installation costs of no more than $10,000 per household, and a useful life span of at least 30 years. The center is working with a number of open-source systems already in use—including constructed wetlands, permeable reactive barriers (described earlier in this chapter), biofilters, and membrane bioreactors—to make refinements and lower costs. The biofilters use a mixture of sand and either sawdust or wood chips to remove nitrogen from the waste stream. Some early laboratory test results have shown nitrogen levels dropping by more than 85 percent. The center is also working on a cellulose membrane technology that uses plants, grass, wood, or other natural materials and that requires less energy than typical systems. While this membrane system is still in the R&D phase, Garvey says it "has the potential to be game-changing" in terms of cost-effectiveness.[15]

In the near term, the best prospect for affordable on-site systems may lie in new technologies developed through private R&D. Several companies have created units that resemble septic systems but use chambers to filter sewage and microbes to convert dangerous compounds into harmless gas. Suffolk County is conducting a demonstration program to showcase a variety of these "innovative and advanced (I/A) treatment units" that have passed the testing required for certification by NSF International, a global public health and safety organization based in Ann Arbor, Michigan.

The technology designed by Adelante Consulting, the one I know best because my spouse is the company's president, is called the Pugo (say aloud for full effect) and has gone through more than a decade of field testing and design iterations. In 2015 field tests, the Pugo reduced total nitrogen by 61 percent, to levels below the targets set by NSF International and Suffolk County.

Advanced on-site treatment units have been around for nearly 50 years, but without regulations that require them or adequate incentives for homeowners to install them, their potential to clean up rivers, lakes, and bays has gone unfulfilled. Some 14,500 advanced wastewater units operate in Massachusetts, Maryland, and Rhode Island, states that began grappling with nitrogen pollution before Long Island did. But the number of I/A units installed in these states, which collectively have hundreds of thousands of septic systems, pales in comparison to the need. On a tour of these regions in 2014, Suffolk County staff learned that getting the costs of I/A systems down is key and that financial incentives are needed for homeowners to adopt them. The state of Maryland, for example, which is working to reduce nitrogen loads to Chesapeake Bay, provides grants for on-site wastewater system upgrades funded in part through special property tax assessments. Massachusetts offers a tax credit for 40 percent of the cost, up to $6,000, for the replacement of septic systems. Even with such financial incentives, however, relatively few I/A systems have been installed, suggesting that some combination of mandates, bigger financial incentives, and lower system costs is needed.[16]

"The politicians are the ones that have to solve this problem," said Garvey. Policies and technologies must work together. "You can't separate the two."

Suffolk County took a big step forward in August 2016 when it changed its sanitary code to give the health department authority to allow the installation of I/A systems, the first big amendment to its code since 1973.

"This was huge," Garvey said. "Now we're in business."

The state is now funding a study to determine how much nitrogen needs to be kept out of Long Island's waters to improve their quality, a scientific determination critical to guiding the establishment of policies and incentives. At a minimum, Garvey says, mandates will be needed for the installation of I/A systems in new construction, when property changes hands, and in priority zones for nitrogen reduction.

Meanwhile, back at the Perelman–Chapman estate on Georgica Pond, those attending the Sunday morning scientific update appeared hungry for good news. Given the net worth of residents living around the pond, the cost of installing advanced treatment units should not be an issue. Nancy Kelley, The Nature Conservancy's Long Island director, announced that her organization's office, which is located in the pond's watershed, would be installing an I/A unit. If enough residents follow suit, Georgica could be an important test case for solutions to harmful algal blooms at an ecosystem scale.

"Rosie did galvanize us to action," said John Hall, referring to his deceased Jack Russell terrier. Ultimately, however, decisions yet to be made will shape Rosie's legacy on the fish-shaped island that now faces a troubling future.

~

When a haunting phrase like *dead zone* appears in news headlines every year like clockwork, something is terribly wrong. But each spring, typically early to mid-June, a press release goes out from NOAA announcing the predicted size of the coming dead zone—a band of water with dangerously low oxygen levels—that will form that summer in the Gulf of Mexico off the southern US coastline. The Gulf's dead zone is the second largest in the global ocean, after that in northern Europe's Baltic Sea. Over the last decade, it has averaged more than 5,600 square miles (14,500 kilometers), an area roughly the size of Connecticut. It typically extends from near shore to about 75 miles (120 kilometers) into the Gulf, and from the Mississippi River west across the Louisiana shelf to Texas.

As on Long Island, an influx of nutrients into the Gulf—mostly nitrogen, but phosphorus, as well—fuels the growth of algae, which then decompose and deplete the water of oxygen. While finfish and other swimmers may migrate to better-quality water, other species may die. As the ecosystem declines, so do the shrimp, crab, and fish harvests and the jobs that depend on them. In 2011 the Gulf's commercial and recreational fisheries yielded a dockside value of $818 million and generated 23 million recreational fishing trips.[17]

Like a giant funnel, the vast watershed of the Mississippi River, which spans 41 percent of the continental United States, collects runoff from as far west as Montana and as far east as Pennsylvania, and moves it through a network of rivers and streams that converge in Louisiana and then empty into the Gulf of Mexico. As with any river that reaches the sea, the Mississippi's delivery of nutrients and sediment from the continental interior to the coast is a natural and important part of the hydrologic cycle; it is essential to the health of the Mississippi Delta and the marine environment. What is neither natural nor healthy is the volume of nutrients now pouring into the Gulf. Over the last half century, as the American heartland transformed from a landscape of native prairies, forests, and wetlands to croplands and cities, water not only began to flow off the land more quickly, increasing the risk of flooding, it began to carry heavy loads of nitrogen and phosphorus, especially from farm fertilizers. Today, agriculture contributes 80 percent of the nitrogen and more than 60 percent of the phosphorus delivered to the Gulf.[18]

In 1997, more than a decade after scientists began to document the scale of the problem, US federal and state agencies formed the Hypoxia Task Force to curb the nutrient load in the Mississippi River basin. In 2001 the task force set a goal of reducing the size of the dead zone to less than 5,000 square kilometers (1,930 square miles) by 2015. Then, early in 2015, the collaborative announced that it would take until 2035—an extra two decades—to achieve that goal. Reaching it would require at least a 45 percent reduction in the loads of both nitrogen and phosphorus flowing into the Gulf.[19]

The failure to make headway is troubling, to say the least. In May 2016, an estimated 146,000 metric tons of nitrate and 20,800 metric tons of phosphorus coursed through the Mississippi River system and into the Gulf, according to data from the US Geological Survey (USGS). The nitrogen load was 12 percent higher and the phosphorus load 25 percent higher than their averages over the period from 1980 to 2015.[20]

Modelers use these May nutrient load estimates to predict the size of that summer's dead zone. Based on the results of several different models, NOAA then releases its official estimate—which for 2016 was 5,898 square miles (15,275 square kilometers), triple the size of the task-force goal. Later in the summer, NOAA deploys a ship to measure the actual size of that year's dead zone, which allows scientists to better understand the dynamics between nutrient loading and certain ocean conditions, such as tropical storms, and to tweak their models for greater accuracy. Due to engine problems with its ship, NOAA canceled the 2016 cruise, the first cancellation in 30 years. But that minor gap in the record does not change the reality.[21]

"We have to continue to focus on nutrient reductions if we are to have healthy and sustainable fisheries," said Nancy Rabalais, a senior scientist with the Louisiana Universities Marine Consortium and a top expert on the Gulf's hypoxic zone. "Unfortunately the long range trend over the past 30 years continues to show little progress towards reducing the dead zone size to the 1,900 square miles that the task force has set."[22]

So what will it take to make progress?

As on Long Island, strategies to keep excess nutrients out of rivers and groundwater must be developed and taken to scale—and in this case the scale encompasses a good portion of the United States. They must involve the right mix of practice and policy, including incentives to get them adopted by property owners, in this case predominantly farmers. And they must be coordinated, sustained, and scientifically monitored to ensure they meet their goals: sustainable agriculture in the American heartland, cleaner water throughout the Mississippi River basin, and a healthy ecosystem in the Gulf of Mexico.

Thanks to the work of agricultural engineers, scientists, conservationists, farmers, and state and federal agencies, the pieces of such a strategy are coming together. Studies show that more judicious and better-timed application of fertilizer, land-treatment methods such as contouring and cover cropping, alternative cropping systems, and the creation and restoration of wetlands, floodplains, and riparian buffers could collectively do a great deal to reduce the volume of nutrients reaching the Gulf. These practices would also safeguard local drinking water supplies, keep rivers and lakes cleaner, and improve habitat for birds and wildlife. In some cases, they actually boost crop yields, while saving farmers money.

The age-old practice of cover cropping, for example, which largely disappeared with the boom in chemical fertilizers after World War II, appears to be making something of a comeback in the American heartland. It involves planting noncash crops, such as clover, sunflower, hairy vetch, and cereal rye in between harvests of corn, soybeans, and other commodity crops. The roots of the cover crops penetrate deep into the soil and hold it in place, while also aerating it, sequestering carbon, and increasing its capacity to hold water. With less soil erosion and water runoff, less nitrogen moves off the land. And with cover crops like clover and hairy vetch that add nitrogen to the soil, farmers can reduce their application of fertilizer, sometimes to zero.

Rulon Enterprises, a family farm business in northeastern Indiana that plants about four-fifths of its acreage with cover crops, estimates that more than half of what it spends on cover crop seed is offset by reduced fertilizer costs, according to reporting by the *New York Times*. Adding in higher yields, less erosion, and other benefits, Rulon estimates the net economic benefits of cover cropping to total $69 per acre. Taking into account the broader societal benefits of improved water quality and the possibility of less nitrogen contributing to the Gulf dead zone would add to the total value.[23]

Despite these benefits, a 2013 study funded by the Arkansas-based Walton Family Foundation and carried out by researchers with the National Wildlife Federation found that cover cropping was used on less than 2 percent of the 277 million acres (112 million hectares) of cropland in the Mississippi River basin. To incentivize greater use of the practice, the state of Maryland, which is outside the Mississippi basin but is working hard to shrink the dead zone in the Chesapeake Bay, reimburses farmers for the cost of cover crop seeds as an effective way to keep more nitrogen on the land and out of the bay. Along with farms in Indiana, those in Maryland rank high in the use of cover crops.[24]

Adjusting the amount of fertilizer and when it is applied can also make a big difference, although, as with cover cropping, farmers may need incentives and technical support to adopt new practices. Many farmers knowingly apply more fertilizer than their crops actually need in order to avoid the risk of lower yields from applying too little. Because crops do not assimilate this "insurance nitrogen," much of it runs off the field to pollute rivers, streams, or groundwater. New techniques such as variable rate application, a practice very similar to the variable rate irrigation described in Chapter 10, uses information technologies and GPS-equipped tractors to match fertilizer applications with soil quality. Research has also shown that much less nitrogen leaves a field if farmers apply fertilizer in the spring rather than the fall.

University of Illinois researchers largely credit more effective use of fertilizer on cornfields for a 10 percent drop in nitrate in the Illinois River from 2010 to 2014 compared to 1980–96. Adding in data for 2015 brought the five-year average nitrate reduction to 15 percent, "a milestone that the state hoped to achieve for all its rivers by 2025," said Greg McIsaac, professor emeritus at the University of Illinois and lead author of a study in the *Journal of Environmental Quality*. McIsaac's research team found that, since 1980, nitrogen fertilizer sales in the watershed held steady while corn yields rose by about 50 percent. That meant more of the nitrogen applied was taken up by the plants and harvested in the grain, leaving less in the soil to run off into the river. Greater Chicago's wastewater district has also reduced nitrogen discharges to the Illinois River, and this may have played a role as well.[25]

The Illinois River—which historically has been a major contributor of nutrients to the Mississippi basin—is one of the few where nitrogen levels now appear to be dropping. A US Geological Survey study found nitrate levels in the river had fallen by 21 percent between 2000 and 2010. The Iowa River, another key source of nutrients, showed a reduction of 10 percent.[26]

But achieving the goal of a dead zone no larger than 5,000 square kilometers (1,930 square miles) requires more measures on more crop-land—and targeting those measures to where nutrient runoff is greatest. A team of researchers led by Sergey S. Rabotyagov at the University of Washington in Seattle found that the goal could be achieved for $2.7 billion per year through cropland conservation and more judicious use of fertilizer in large portions of the upper Mississippi River basin and the Ohio–Tennessee River basins, with additional investments in a number of the other sub-basins. Doing the same throughout the whole watershed could shrink the dead zone to about 2,900 square kilometers (1,120 square miles) but would cost some $5.6 billion per year, twice as much as the more targeted strategy.[27]

Rabotyagov's team considered only measures that wouldn't require farmers to change what they planted or how much they produced. Nitrogen would drop further if the farmers planted cover crops, shifted cropping systems, planted buffer strips between their fields and nearby streams, and converted some cropland back to native grasses and wet-lands.

In the upper Mississippi watershed alone, some 35 million acres (14 million hectares) of wetlands have been drained. That's an area the size of Illinois that can no longer filter nitrates and other would-be pollut-ants. Much like the microbes in the advanced on-site sewage treatment units, microorganisms in wetland soils convert nitrogen compounds into harmless gases. As farms replace wetlands in the Mississippi basin, this natural service disappears, even as the need for it rises with fertilizer use. One way to reverse this trend is to construct wetlands that would treat runoff before it pollutes rivers and streams; the wetlands would also capture and store floodwaters during storms.[28]

In 2015, biogeochemist Mark David at the University of Illinois found out just how effective wetlands might be at this task. His team conducted a two-year study on wetlands constructed 20 years earlier between tile-drained farm fields and the Embarras River, an Illinois tributary of the Wabash River. They found that the wetlands were taking 62 percent of the nitrates out of the fertilizer-laden farm runoff. As an added bonus, there were minimal emissions of nitrous oxide, a potent greenhouse gas.[29]

"By building a wetland, farmers have an opportunity to make a substantial reduction in the transport of nitrate from their fields to the Gulf," David said in a news release on the study.[30]

Iowa's Conservation Reserve Enhancement Program, a federal, state, and local partnership, has similarly found that strategically placed wetlands remove 40 to 70 percent of nitrates (and over 90 percent of herbicides) from farm drainage. Research by The Nature Conservancy suggests that a wetland area equal to about 6 percent of the tiled farm field removes about 50 percent of the nitrates draining off the field. The Nature Conservancy and its partners are working with Illinois farmers to build wetlands in places that effectively trap nitrates before they pollute the Mackinaw River, a major tributary of the Illinois River and a source of drinking water for more than 80,000 people.[31]

Taken together, these measures to restore wetlands and riparian forests, reduce soil erosion, and apply fertilizer in the proper amounts and at the right times start to add up. Reducing nitrates by 45 percent basinwide doesn't feel so out of reach. But there's one big hitch: without adequate incentives and support, few farmers are voluntarily going to change their practices or give up a portion of their fields for wetlands or buffer strips. As economists would say, the external costs of row-crop agriculture in the Mississippi basin—namely, nitrogen pollution and the Gulf dead zone—must be internalized.

Most farmers care for their land and want to steward it properly, but like any business, they need a financial reason to do so. A tax on fertilizer that makes its price better reflect its true cost would be a good start. So would even a limited reform of the farm bill. A 2006 analysis by the Washington, DC–based Environmental Working Group found that taxpayer-funded federal crop subsidies in the Mississippi basin exceed spending for conservation measures to improve water quality by more than 500 to 1.[32]

Shifting a modest portion of those crop subsidies—a good share of which go to wealthy agribusiness—into programs that encourage nitrogen reduction could substantially shrink the dead zone, while mitigating floods, protecting drinking water, and expanding wildlife habitat at the same time. The Environmental Working Group study found that many farmers want to participate in these conservation programs but cannot because the programs are underfunded. In 2004, some 2,450 farmers who tried to enroll 321,000 acres (130,000 hectares) in the federal Wetlands Reserve Program were turned down because there was no money.

Agricultural subsidies involve powerful interests and tricky politics. But if the nation wants to protect its premier fishery in the Gulf, political leaders of all stripes will need to step up to the plate. Taxpayers who think it's wrong that for every federal dollar that supports industrial agriculture less than a penny goes to cleaning up its pollution might call or write their representative.

Just as too much cholesterol or sugar in our bloodstream threatens our overall health, too much nitrogen coursing through the water cycle threatens the whole ecosystem. From farmlands to septic systems, it is possible, and it is time, to clean it up.

CHAPTER 9

Close the Loop

In Nature, there is no waste.

Aʙᴏᴜᴛ sɪx ʏᴇᴀʀs ɪɴᴛᴏ what would become known as the Millen-nium Drought in Australia, the managers of an attractive golf club, Pen-nant Hills in the state of New South Wales, grew anxious. Reservoir levels around Sydney, the state's capital, had dropped to record lows. To stretch the city's dwindling supply, Sydney Water tightened restrictions on water use. The club, founded in 1923 and boasting a championship course, would get no more than 20,000 cubic meters (5.3 million gal-lons) of water per month, well short of the amount it normally used for watering. The club managers pictured their prized greens turning to ugly browns. But rather than cross their fingers and hope for the best, they took an unusual step: they requested permission to tap into the sewer line that ran beneath the golf course. The club's plan was to treat that sewage on-site and then use it to irrigate its 23 hectares (57 acres) of greens.

It might sound yucky, but "sewer mining," as the Aussies call it, is catching on, not just in Australia but also around the world. As the name implies, it involves tapping into a wastewater collection system and siphoning off some of the sewage that's moving through the pipes. The sewage then gets treated and used for landscape irrigation, toilet flushing, and other nonpotable uses.

Many of the world's dry regions now treat and reuse their wastewater to drought-proof or augment their supplies. In 1968 Windhoek, the capital of the southern African nation of Namibia, startled the world by becoming the first city to reclaim its wastewater with advanced treatment processes and then drink it. With annual rainfall of 370 millimeters (14 inches) and no perennial river within 750 kilometers (466 miles), the city had few options other than to close the loop on its water system. Initially skeptical, the residents of Windhoek have been drinking their reclaimed wastewater for nearly half a century with no outbreaks of disease or negative health effects.[1]

Many other cities have followed the pioneering steps of Windhoek, but most add an additional step: instead of drinking their reclaimed wastewater directly, they first send it to an aquifer or surface reservoir to dilute it and allow for some additional natural treatment before sending it through their drinking water treatment plant and supplying it to their residents. This extra buffer has generally made what water managers call "indirect potable reuse" more acceptable to consumers than the "direct potable reuse" (sometimes unhelpfully called "toilet to tap") practiced in Windhoek.

Orange County in southern California now runs the world's largest indirect potable reuse operation. Through processes such as microfiltration, which removes almost all suspended solids, bacteria, and protozoa, followed by reverse osmosis, which forces the water through a membrane and removes viruses, pharmaceuticals, and other impurities, followed by a round of disinfection, Orange County treats its sewage to drinking-quality levels. It then stores the reclaimed water underground for a period of time before pumping it up and delivering it to residents. Along with conservation and storm-water capture, the reuse of wastewater is helping wean southern California from its dependence on long-distance water from the Colorado River and the northern part of the state.[2]

Sewer mining differs, however, in that it is decentralized and localized. Instead of collecting a whole city's wastewater, sending it to a large treatment plant, and then piping the reclaimed water great distances for reuse, sewer mining allows for the treatment and reuse of wastewater in or near the same location. "Small-scale, modular, localized wastewater treatment is now becoming part of our infrastructure," says Stuart White, director of the Institute for Sustainable Futures at the University of Technology Sydney. White sees such schemes as examples of water infrastructure's "fourth generation."[3]

This decentralized approach offers a variety of potential benefits. It can relieve overtaxed wastewater systems, trim the costs of building and maintaining treatment plants and piping networks, reduce energy and chemical use, and save drinking-quality water for actual drinking. These savings, in turn, keep more water in rivers, lakes, and streams—which is especially crucial during droughts and summer months, when river flows are low and water demands are high.

"Sydney Water initially couldn't get its head around the concept (of sewer mining)," said Kurt Dahl, managing director of Permeate Partners, the consultancy that helps operate the Pennant Hills system. I toured the operation with Dahl when I visited the golf club in June 2010.[4]

The sewer pipe running through the golf course carries wastewater from about 1,000 homes to the coastal town of Manly, some 15 kilometers (9.3 miles) northeast of Sydney. There the sewage receives only very basic treatment before being dumped into the ocean. So Pennant

Hills was tapping wastewater that would otherwise not only go unused but also pollute the South Pacific. As long as the golf club siphoned off its new supply during peak hours of toilet flushing and showering—the morning and evening—it would not interfere with the pressure and flow rate needed to get the remaining sewage to Manly. The little treatment plant sits unobtrusively adjacent to the tenth fairway, surrounded by trees and gardens. It produces virtually no odor.

The sewer-mining scheme has cut Pennant Hills's potable water use by 92 percent, which earned the club an award from Sydney Water. As a bonus, nitrogen in the sewage is transformed from a harmful pollutant into a valuable nutrient. Pennant Hills has nearly eliminated its use of chemical fertilizers because small amounts of nitrogen get added to the greens each time they are irrigated. Overall the system has proven to be a cost-effective way to drought-proof the links and reduce stresses on Sydney's water supply. Even the golfers are pleased. "Old-time club members say this is the best the golf course has looked in 30 years," Dahl said.

What makes sewer mining feasible, and the reason it seems likely to catch on and spread, are advancements in a variety of treatment technologies over the last decade. At the heart of the Pennant Hills system, which was designed by GE Water, a division of General Electric, is a technology called a membrane bioreactor (MBR). After microorganisms treat the sewage biologically, the resulting product gets drawn through a membrane with microscopic pores that let the partially treated water through but block almost everything else. The waste sludge, which constitutes about 2 percent of the original sewage, then returns to the sewer, while the treated wastewater gets disinfected with chlorine or ultraviolet radiation before being sprinkled onto the gardens and greens.[5]

This MBR process has been in use for several decades, but in recent years both the cost and energy requirements of the membranes have declined substantially. As a result, MBRs produce higher-quality water and cost roughly the same as other technologies. MBRs treat more than 3 billion liters (792 million gallons) of water a day, and installed capacity is growing rapidly. (For more on nitrogen-reducing technologies, see Chapter 8.)[6]

Dockside Green, a 15-acre (6-hectare) mixed-use development in Victoria, British Columbia, and one of the first planned communities to earn Platinum certification for Leadership in Energy and Environmental Design (LEED), uses an MBR system to treat all of its sewage. The reclaimed water is used to flush toilets, irrigate the landscape, and add flow to a local creek. According to Chris Allen, regional manager with General Electric's Water and Process Technologies division, sewer mining, along with conservation measures such as dual-flush toilets, water-efficient fixtures, and gray water systems, have reduced indoor water use by 65 percent. Solaire, a 293-unit apartment complex in the Battery Park neighborhood of New York City, is the first onsite water reclamation system in the United States built right inside a residential apartment building. According to Allen, it recycles about 25,000 gallons (94,635 liters) per day to the building's cooling towers, toilets, and landscapes.[7]

One of the world's largest sewer mining projects is at Cauley Creek, a high-end community in Fulton County, Georgia, northeast of Atlanta. As with Sydney's Pennant Hills Golf Club, Cauley Creek residents became concerned that drought would lead to restrictions on water withdrawals from the Chattahoochee River, the source of their drinking water. The community's MBR system can reclaim 5 million gallons (roughly 19 million liters) of wastewater per day, which gets used by area schools, churches, homes, and a golf course. The on-site treatment and reuse allow more water to remain in the Chattahoochee, a river with an exceptionally high diversity of fish, mussels, and other aquatic life. In keeping with the community's rural character, the treatment plant sits quietly inside a classic-looking red barn, complete with a weather vane.

Like Sydney, many coastal cities around the world discharge some or all of their wastewater, treated or not, into the ocean. Shanghai, China's most populous city, sends wastewater to the Pacific. So does Victoria on Canada's Vancouver Island. And, as the media spotlighted in the run-up to the 2016 Summer Olympics, much of the sewage from the 12 million inhabitants of coastal Rio de Janeiro, Brazil, gets dumped into local waters without adequate treatment. US cities and towns collectively discharge 32 billion gallons (121 billion liters) of wastewater every day, and about 12 billion of those gallons (45 billion liters) get treated and released to an ocean or estuary. Each day that's like throwing away a volume of water equal to 12 times the daily water use of New York City.[8]

Consider Boston, Massachusetts. In the 1980s, when a judge ordered the state capital to stop dumping raw sewage into Boston Harbor, the municipality decided to build a spanking new wastewater treatment facility on Deer Island, a peninsula that extends into the harbor. Today, sewage from 43 greater Boston communities gets transferred to Deer Island, where it gets treated and then sent through a 9-mile (14.5-kilometer) tunnel that discharges into Massachusetts Bay. Once swallowed by the salty sea, the treated water can be of no further service to the state's communities or freshwater ecosystems.

As water supplies tighten, rivers dry up, wetlands shrink, and persistent droughts lead to water cutbacks, cities and farms are taking a new look at wastewater. It is increasingly viewed not as a nuisance but as an asset. For many activities—from toilet flushing to landscape watering to crop irrigation—water does not need to be clean enough to drink. Matching water's quality to its intended use opens up a whole new way of thinking about and managing water, one that can turn wastewater from a disposal problem into a valuable new source of supply.

If the whole idea of using wastewater raises eyebrows, it helps to remember that all water on Earth is recycled. More to the point, many cities live downstream from others, so one city's wastewater discharge becomes another's drinking water supply. Europe's Danube River carries wastewater releases from Vienna, Austria, to Bratislava, Slovakia, those from Bratislava to Budapest, Hungary, and those from Budapest to Belgrade, Serbia. As long as each city treats its wastewater before releasing it to the Danube, and also treats and disinfects any river water it extracts for its residents to drink, water quality and public health can be protected.

This kind of unplanned or de facto reuse happens all the time. Residents of New Orleans drink water from the Mississippi River that has passed through the water and sewer systems of a dozen cities upstream. About half of the water flowing into the main reservoir for Houston, Texas, called Lake Livingston, is treated wastewater from the Dallas–Fort Worth area. That wastewater flows for about two weeks in the Trinity River, where natural microbial and sun-driven chemical processes boost its quality further before it enters Houston's reservoir. There it undergoes still more natural cleansing before engineers run it through the city's water treatment plant to remove remaining impurities. After disinfection it is supplied to Houston residents for drinking, cooking, and other household uses. During dry times, those wastewater flows from Dallas–Fort Worth make up most of the Trinity River's flow; without them, the river would nearly stop flowing.[9]

More and more, cities are incorporating the reuse of wastewater into their long-term planning. In some cases, as in Windhoek and Orange County, the wastewater is treated to such a high quality that it meets drinking water standards. In others, the reclaimed wastewater serves nonpotable uses, such as irrigating lawns, parks, golf courses, and farms; replenishing groundwater; or creating wetland habitat. California, a US leader in the practice, is estimated to reuse 670,000 acre-feet (826 million cubic meters) of municipal wastewater each year, roughly equal to the water used by 1.4 million households.[10]

Many industries are also turning to reclaimed wastewater as a secure, sustainable supply. Some treat and recycle their own process water, while others purchase reclaimed wastewater from a local utility. In June 2016, Apple, the maker of computers and mobile phones, announced that it would pay for a new wastewater treatment facility in the central Oregon town of Prineville and use the reclaimed water to cool its data centers there. By using recycled water, Apple will save Prineville some 5 million gallons (19 million liters) a year.[11]

∼

Israel was among the first nations to have an "aha" moment about the potential of urban wastewater: recycling it to farms could alleviate water shortages and bolster food security. In dry regions, irrigated agriculture is nearly always the biggest user of water. Globally, it accounts for 70 percent of total water withdrawals from rivers, lakes, and aquifers. Typically half or more of the water applied to farms is "consumed" through transpiration and evaporation. The inherent thirstiness of crops combined with inefficient irrigation is a major reason so much of the world faces water shortages.

Within a decade of Israel's founding in 1948, the new nation's officials, scientists, and engineers began designing a scheme to collect the wastewater generated by cities and towns in the greater Tel Aviv area and deliver it to a centralized treatment facility 13 kilometers (8 miles) south of the city. Like many coastal cities, Tel Aviv had been discharging its wastewater to the sea, in this case the Mediterranean. The treatment plant was finally completed in 1973. Meanwhile, as recounted by Seth Siegel in *Let There Be Water*, an agriculture official pitched the idea of transporting the treated Tel Aviv wastewater to farmers in the Negev Desert who needed a secure supply of water to irrigate their crops. After tests showed that the natural filtration provided by the fine sands overlying the aquifers in the treatment area would remove any remaining impurities of concern, the path was clear to send farmers their new water supply. Engineers built a pipeline 1.8 meters (6 feet) in diameter and 80 kilometers (50 miles) long to transport Tel Aviv's reclaimed wastewater to farms in the Negev. Today, Israel reuses over 85 percent of its municipal wastewater, and the recycled water meets nearly half of the nation's irrigation demand.[12]

The application of wastewater to farmland has a long, storied, and somewhat controversial history. Early on, many European cities simply sent their raw sewage to farmlands or pasture, counting on the land to naturally filter impurities. Such "sewage farms" were operating in Edinburgh, Scotland, as early as 1650, and soon after sprang up outside of London, Manchester, and other English cities. In 1897, Melbourne, the capital of the Australian state of Victoria, began sending its sewage to Werribbee Farm, about 30 kilometers (18.6 miles) to the west. Using a combination of simple treatment lagoons and land-filtration methods, Werribbee irrigated nearly 11,000 hectares (27,000 acres) with Melbourne's sewage, the largest system of its kind in the world. To help manage the lush grasses that grew abundantly from the nitrogen and phosphorus loads in the wastewater, sheep and cattle grazed the irrigated pastures.[13]

While this scheme worked reasonably well for nearly a century, scientists then started to detect high levels of nitrogen in Port Phillip Bay, on the coast near Werribbee. In 2004, in response to growing concerns about pollution, Melbourne Water switched from the basic lagoon and land-treatment processes to modern waste stabilization ponds capable of delivering a higher-quality effluent. Today, Melbourne Water's operation at Werribbee treats just over half of the city's domestic sewage. While sheep still graze irrigated pastures at Werribbee, the land is now used for a mix of crops, pasture, horticulture, and forestry. The wetlands at Werribbee are one of Australia's most important sites for waterbirds, as well as prime habitat for the endangered growling grass frog. They are listed as wetlands of international importance under the Ramsar Convention.[14]

Early experiments like those at Werribbee helped pave the way for beneficial partnerships between cities and farms around wastewater recycling and reuse. A surprisingly large share of the world's cropland is found in and around cities. Some 67 million hectares (166 million acres)—including 11 percent of the world's irrigated land—is located within an urban core, according to a study by the International Water Management Institute in Colombo, Sri Lanka. Adding in land cultivated within 20 kilometers (12.4 miles) of an urban perimeter brings the total area of urban and peri-urban farms to 456 million hectares (1.13 billion acres). That's an area roughly the size of the European Union, and it does not include the small backyard plots and rooftop gardens from which a growing number of city dwellers harvest vegetables and herbs. Because urban farms are typically managed quite intensively, they often yield more food per hectare than farms in the countryside do.[15]

For regional water managers, recycling treated municipal wastewater to farms close to the towns that generate it can be a cost-effective way of getting double duty out of their water supply. Using water twice can also alleviate pressures on rivers and aquifers. For farmers, the nitrogen and phosphorus contained in the wastewater are valuable nutrients that can reduce the use and cost of chemical fertilizers. As long as the treated wastewater meets health and environmental standards, and its distribution to farms doesn't deprive or harm a river or other water source, it can be a winning proposition.

That has more or less been the experience of farmers and water managers in central California's Salinas Valley, the "salad bowl" of the United States. Back in the 1970s, the region encountered a serious problem: its groundwater was getting salty. Decades of heavy pumping from the region's aquifer had caused seawater from Monterey Bay and the Pacific to move inland, contaminating the underground freshwater supply. To combat this invasion of saltwater, the community decided to explore the idea of replenishing the aquifer by allowing farmers to switch their irrigation supply from groundwater to treated municipal wastewater. While the use of reclaimed wastewater to irrigate landscapes and certain crops was by then well accepted, the idea of applying it to edible fruits and vegetables, including celery, lettuce, strawberries, and other crops eaten raw, was a new idea that understandably aroused some concerns. The growers were game to give it a try, but while the scheme met the state's regulatory requirements, the county's health official needed further convincing that it would not jeopardize the public's health.[16]

In response, the county launched a major research effort to evaluate the risks, and, in an astute move, made the cautious county health officer the chair of the research committee. Over 11 years, from 1976 to 1987, the committee studied contamination risks from viruses, bacteria, and heavy metals, as well as worker safety and consumer health issues. In the course of the study, which other countries would later turn to as something of a model, nothing of concern turned up. The health official "got all his questions answered," recalled Keith Israel, former general manager of the Monterey Regional Water Pollution Control Agency. "He went all the way from being the biggest skeptic of the project to the biggest supporter."[17]

The Salinas Valley Reclamation Project, which came online in 1997, is the largest sewage treatment facility in the world that recycles wastewater for irrigation of edible food crops. It treats up to 29.6 million gallons (112 million liters) of wastewater per day to advanced (tertiary) levels for distribution to 12,000 acres (4,860 hectares) of valley farmland in northern Monterey County. Groundwater levels in the valley have risen, and seawater intrusion has slowed. The organic nitrogen in the wastewater now fertilizes the land instead of polluting the bay.

These benefits notwithstanding, more research is needed to understand the long-term effects of using wastewater to irrigate crops on different types of soil. Although the constituents of wastewater vary depending on the quality of the raw sewage and the level of treatment it is given, in general wastewater contains higher loads of dissolved organic matter, salts, and suspended solids. Researchers have found that, after being irrigated with wastewater for an extended period, citrus and avocado trees grow more slowly, yield less fruit, and may take up less water. Similarly, scientists in Israel found that 15 years of irrigation with treated wastewater changed clayey soils in important ways, including how much water they could retain.[18]

It's difficult to get a handle on how much wastewater is actually used for irrigation in the world today, in large part because, out of necessity, farmers in poor countries often irrigate with whatever water is available, even untreated wastewater. In much of Asia and sub-Saharan Africa, less than 10 percent of the urban population is connected to a piped sewer system. That means raw sewage flows through urban areas in ditches, canals, and rivers. For poor farmers in need of irrigation water, untreated sewage, while posing health risks, is a precious supply. It is available year-round and contains valuable nutrients, enabling them to generate higher yields and incomes. Even when the use of raw sewage to irrigate crops is officially prohibited, the practice may be "unofficially tolerated" because farmers' livelihoods depend on it.[19]

In the West African nation of Ghana, for example, some 2,000 urban farmers supply about 800,000 residents with vegetables that are often irrigated with untreated sewage or river water polluted by it. As much as 10 percent of household wastewater in the capital city of Accra is used untreated on urban vegetable farms. "These farms are now recycling more wastewater than local treatment plants," according to Pay Drechsel, an environmental scientist with the International Water Management Institute (IWMI) in Colombo, Sri Lanka, and a recognized leader in the reuse of wastewater in agriculture. By some estimates, about 10 percent of global crop production comes from farms irrigated with wastewater receiving little or no treatment.[20]

In an official nod to these realities, the World Health Organization (WHO) in 2006 revised its guidelines for the reuse of wastewater in agriculture. Rather than let the perfect be the enemy of the good, the WHO called for a determination of how much additional disease could be tolerated as well as a multiple barrier approach to reducing the risks of exposure to bacteria. Those barriers may include low-cost, on-farm treatment systems, such as wastewater ponds and simple filtration methods. Researchers at IWMI and elsewhere are designing and testing a variety of on-farm systems for practicality, affordability, and effectiveness. While these systems do not treat wastewater to the levels required in wealthier countries, when combined with farmer and consumer education, they can reduce the risk of disease while providing farmers an irrigation supply that allows them to grow nutritious vegetables and increase their incomes.[21]

~

One final note about the reuse of wastewater: it can play an instrumental role in restoring wetlands for birds and wildlife. Not only do birds flock to the marshes and lagoons, as witnessed at the Werribbee treatment facility outside of Melbourne, but the wetlands further treat or "polish" the wastewater before it is discharged to local rivers or streams.

Constructed or engineered wetlands use the same interactions of plants, soils, microbes, water, and sunlight that natural wetlands do to remove pollutants and cleanse water supplies. Communities across Europe and North America have been constructing wetland treatment systems for decades to improve the quality of their rivers, lakes, and aquifers. If located in the right place, constructed wetlands can also provide homes for birds and wildlife where their natural habitats have largely disappeared.

This is exactly what is happening in Las Arenitas, a sewage treatment facility in the Mexican state of Baja California. There, urban wastewater that once made a smelly health hazard of the New River, which flows north into California from the US–Mexico border, is now sustaining a wondrous wetland and bird-watchers' paradise in the northwest corner of the Colorado River Delta. As described in Chapter 2, the delta has lost about 90 percent of its wetlands as water has been siphoned off upstream. That has left resident birds and wildlife, as well as birds migrating along the Pacific Flyway, with few places to rest, nest, and feed.

Located about 16 miles (26 kilometers) southeast of Mexicali, Las Arenitas looks at first glance like any other municipal wastewater treatment plant. An underground pipeline daily delivers nearly 20 million gallons (76 million liters) of Mexicali's sewage to the facility, where it then undergoes conventional physical, biological, and disinfection processes. Originally the treated effluent was discharged into a drain that emptied into the Hardy River, which flows south and joins the Colorado River shortly before it reaches (or used to reach) the upper Gulf of California. Unfortunately, the plant's initial startup in 2007 did not go well. Not only did it fail to meet the required water quality standards, it sent raw sewage flowing down the Hardy. The indigenous Cucapá living along the river were understandably irate.[22]

A year later, as the public services commission in Mexicali was designing improvements to the Las Arenitas plant, two conservation groups —Pronatura Noroeste, based in Ensenada, Baja California, and the Sonoran Institute in Tucson, Arizona—seized upon an opportunity. The groups proposed constructing a wetland at the site that would provide additional treatment to the wastewater and restore vital bird habitat. Javier Orduño Valdez, formerly a congressman from Baja California who became director of the public services commission, signed on to the idea. The conservationists and the commission partnered to construct a 250-acre (100-hectare) marsh of cattails and native vegetation.

During my visit to Las Arenitas in February 2013, the wetland was a riot of birds. A yellow-rumped warbler sang from atop a mesquite tree. Flotillas of American coots plied the lagoons. Hiding in the cattails, a secretive sora delivered its telltale descending whinny. The first bird count at the marsh, in 2009, yielded eight species; within 6 years, there were 160 species. Even the endangered Yuma clapper rail is using the site to breed. The Hardy River is flowing healthier too: the treated wastewater from the marsh has roughly doubled the river's flow, creating better fishing and economic opportunities downstream.[23]

For Orduño Valdez, Las Arenitas anchors a vision for recreation, tourism, and economic development in the region. He described plans to build a nature park with trails around the marshes, along with exhibits to explain the importance of the constructed wetlands to the Rio Hardy and to the restoration of the Colorado Delta. "With this water we give life to all this area," he said.

In a world of mounting competition for water, reuse is a crucial part of the solution. It can help grow food and boost economies, while reducing pollution in rivers and bays, the use of chemical fertilizers, long-distance food and water imports, and related greenhouse gas emissions. The use of wastewater requires balancing concerns about human health, environmental quality, and food and water security. But looking ahead, closing the loop will be less a choice, and more a necessity.

CHAPTER 10

Let It Flow

Right now, in the amazing moment that to us counts as
the present, we are deciding, without quite meaning to,
which evolutionary pathways will remain open and which
will forever be closed.
—*Elizabeth Kolbert*

ON A VISIT BACK TO MY OLD STOMPING GROUNDS in western Massachusetts in early January 2014, I trekked along the snowy banks of Amethyst Brook, a beautiful tributary in the watershed of the Connecticut River, the largest river in New England. As a longtime resident of Amherst, I'd hiked through this conservation area many times, mostly to enjoy the woods, the babbling brook, and the birds. But on this wintry day, I had a mission: to check out the site of an early nineteenth-century dam that had been demolished a little over a year before.

The rock structure, 20 feet (6 meters) tall and 170 feet (52 meters) wide, had blocked the flow of Amethyst Brook since 1820. It was called the Bartlett Rod Shop Company Dam, after the maker of fly-fishing rods that had started production at the site in 1864. Fishermen relished the split-bamboo rods, but ironically the dam had blocked the upstream movement of the native brook trout they prized. As I scouted around that morning, I saw the remnants of the dam on both sides of the brook, which now flowed freely toward the Fort River, the next tributary downstream. In addition to the fishing-rod maker, power from the Bartlett Dam had serviced a woodworking shop, a machine shop, and a producer of cleaning equipment for boiler tubes.

As happens over time, the dam fell into disrepair. In 2007, state officials issued a dam-safety order to the current owner of the site, HRD Press, which had no use for the relic structure. When confronted with the choice of either repairing the dam for about $300,000 (plus maintenance in the future) or taking it down for $193,000, the company quickly settled on dismantlement. Then in 2010 state officials sweetened the deal by declaring the Bartlett Dam's removal a priority for river restoration. A broad coalition of federal and state agencies, the towns of Amherst and Pelham, and local and national conservation groups got busy developing a plan.

By late 2012, the dam was gone. Almost immediately, the little river began to heal. Sediment trapped behind the dam flowed downstream, along with leaves, twigs, and other debris critical to aquatic food webs. Water temperature dropped. Oxygen levels rose. Within six months, a snakelike fish known as the migratory sea lamprey was spawning in gravel and cobbles in the stream bottom just below the old dam site. The newly created habitat was just right for the boneless native to lay her eggs. It was "a great sign of improving conditions in the river," noted Amy Singler of American Rivers, one of the key partners in the dam's dismantling.[1]

The unshackling of Amethyst Brook is part of a growing movement of dam removals across the United States. Especially in the East, mill dams dating to the early years of the industrial revolution, now obsolete and often hazardous, are coming down. "If it was just habitat restoration alone, we wouldn't be doing so many dam-removal projects," said Brian Graber of American Rivers, when I met with him and Singler shortly after my visit to Amethyst Brook. Throughout New England, 80 dams have been removed over the last 15 years, with an additional 65 removals planned.[2]

Removing a dam is the simplest and most effective way to restore river habitat and open up corridors for fish. Just as a blocked artery prevents blood and oxygen from reaching vital organs in the body, dams prevent rivers from delivering the flow, sediment, and nutrients needed to keep the river system functioning. Without the dam, a river reclaims some of its "natural flow regime," the unique pattern of highs and lows across the seasons and years that is essential to the river's health, and to which fish and other creatures have become adapted over time.[3]

Across the United States, some 76,000 dams at least 6 feet (1.8 meters) high block rivers and streams.[4] Many fish populations have plummeted as a result. In North America, only 135 of some 600 original runs of Atlantic salmon remain. American shad have disappeared from nearly half of the rivers in which they once swam. In his exquisite book *Running Silver*, conservation biologist John Waldman refers to these lost populations as "ghost fishes," because their absence leaves holes in the ecological web.[5] Freshwater mussels, for example, rely on certain fish to carry their larvae for several weeks before the juvenile mussels drop from the host fish and burrow into the river's sediment to mature. Alewife, American shad, and blueback herring serve as hosts for a mussel called the alewife floater. If those fish species disappear, the alewife floater will likely vanish too.[6]

During the last 30 years, partnerships of cities and towns, state and federal agencies, and conservation organizations have removed 1,174 dams from rivers and streams across the United States, according to American Rivers. While most are relatively small, like Bartlett, a few are big dams, like Glines Canyon and Elwha, which were dismantled between 2011 and 2014 on the Elwha River in Washington state. Within months of the first dam's removal, scientists counted more than 4,000 spawning Chinook salmon above the former dam site. Today fish populations in the Elwha are at levels not seen in 30 years.[7]

Similarly, for more than a century, four hydropower dams on Maine's Penobscot River blocked the upstream migrations of a dozen fish species, including one of the largest runs of Atlantic salmon in the United States. An extraordinary collaboration between the Penobscot Indian Nation, two hydropower companies, seven conservation organizations, and state and federal agencies resulted in the two lower dams, Great Works and Veazie, coming down in 2012 and 2013. Along with the bypassing of the third dam, these removals reconnected 1,000 miles (1,600 kilometers) of freshwater habitat, benefiting salmon, shad, alewife, and other fish species that historically migrated from the Atlantic up the Penobscot. For the first time in over a century, female shortnose sturgeon, an endangered species that has plied Earth's waters for millions of years, are reaching habitat in the Penobscot suitable for them to reproduce.[8]

The Penobscot partnership also shows that removing a large dam does not have to mean a drop in energy production. Equipment upgrades and improved operations at other Penobscot dams allow the basin to produce the same amount of power, even as the river and its fish populations bounce back to health.

Of course, while removing a dam is good for a river, not building one in the first place is better. While countries in Asia, Africa, and Latin America continue to dam their rivers to generate electricity as the United States, Canada, and European nations did throughout the twentieth century, the pace of construction has slowed. Wind and solar options have become more cost-effective and pose far fewer ecological and social concerns. Big water projects are risky business, particularly as climate change creates more extreme floods and droughts. According to reporting by Circle of Blue, wind and solar projects that came online in 2015 have the ability to produce five times the electricity of new hydropower projects.[9]

The age of big dams is by no means over, but some countries and investors are pivoting away from big hydro toward more sensible options. In early 2016, Cambodia declared a moratorium on the construction of large hydropower dams until at least 2020, a welcome reprieve for the biologically rich Mekong River system and the millions of people who rely on its fisheries for their protein. In the Indian state of Assam, citizen opposition in 2011 forced the shutdown of the Lower Subansiri hydropower project. And Brazil's environment agency has reportedly suspended the construction of what would be the South American nation's second-largest hydropower scheme.[10]

The fact remains, however, that most of the world's large dams (those at least 15 meters, or 49 feet, high)—which now number more than 58,500—will remain standing for the foreseeable future, and many of the world's rivers will continue to be turned on and off like plumbing works.[11] But coalitions of scientists, communities, government agencies, and conservationists are finding ways to give rivers the flows they need to regain some health and function. As we'll see, these efforts can take many forms—including operating dams differently, setting limits on water diversions from rivers at risk, and irrigating more efficiently to keep the saved water in-stream. None of these are easy to do, and bringing these efforts to scale will take time, incentives, policy reform, and funding. But a blueprint for healthier rivers starts to take shape.

The dismantling of Bartlett Dam on Amethyst Brook in western Massachusetts is one small piece of a big vision for restoration of the Connecticut River watershed. The basin covers an area nearly the size of Maryland and extends from the New Hampshire–Quebec border to the

Long Island Sound. For a dozen years, Kim Lutz, who directs the Connecticut River Program for The Nature Conservancy (TNC), has spearheaded a wide-ranging restoration effort in partnership with university scientists, state and federal agencies, and local communities. The undertaking is daunting. The basin includes more than 3,000 dams across a network of 20,000 river miles—one of the highest dam densities of any watershed in North America.

The Connecticut River basin in the northeastern United States.

"We had to build hydrologic models from scratch," Lutz said in her office in Northampton, Massachusetts, where we met up in late June 2016. Because the data from river gauges go back 100 years while many of the basin's dams date back over 200 years, the task of reconstructing the river system's historical flow pattern—a first order of business—was difficult, to say the least. Prior to the models, "it was anyone's guess," Lutz said. "We now have information to make good decisions, and we've added to the body of science."

The body of science Lutz refers to has been built up over the last two decades by pioneers in Australia, the European Union, South Africa, and the United States who focused on a critical question: Can we manage dams in ways that make rivers healthier without sacrificing the benefits of flood control, hydroelectric power, recreation, and water supply? The concept of "environmental flows" and the "natural flow regime" entered the lexicon and became guideposts for freshwater conservationists in many parts of the world.

Then in 2002, TNC partnered with the US Army Corps of Engineers (the Corps) to see if these concepts could work on the ground. Under the leadership of Brian Richter, TNC had developed methods for determining the pattern of flows required to keep a river healthy. The Corps, which operates 692 large dams across the United States, was an ideal partner for testing these methods. A new initiative called the Sustainable Rivers Project (SRP) was born.[12]

The project that got TNC and the Corps hooked was on Kentucky's Green River, a beautiful tributary of the Ohio River that harbors one of the most diverse assemblages of fish and freshwater mussels in the United States. The team used information about the river's natural flow and the habitat of particular species to model different scenarios of dam operations. They discovered that slightly reducing the reservoir's flood pool, and adjusting when the reservoir is drawn down for the winter and refilled in the spring, would benefit the ecosystem without increasing flood risks. Moreover, the longer recreation season at the reservoir would boost revenues from boating and other activities by 45 percent. Importantly, the Corps was able to make the change: it was clear and operational. In 2006, after a three-year trial, the Corps revised its guidance for the reservoir to lock in these benefits to the river and the local economy.[13]

By 2012, the SRP and its partners had released river-healing flows at 11 dams on six rivers. On Arizona's Bill Williams River, a tributary of the Colorado, flows released at the Alamo Dam have revived the river's floodplain forests, a boon to more than 350 species of birds and hundreds of thousands of bird enthusiasts.[14] In the Caddo Lake–Big Cypress watershed, a lush wetland ecosystem that spans the Texas–Louisiana border, the Army Corps and TNC are working with partners to bring back populations of American paddlefish—an ancient fish closely related to sturgeons whose populations have plummeted over the last century.[15]

"Today we have good science that can be used to modernize the operations of dams," says Andy Warner, who spent a decade coordinating the SRP at TNC and is now a fellow with the Corps's Institute for Water Resources. Warner hopes that the experience gained through the SRP will eventually lead to real changes at as many as 600 dams, benefiting 50,000 miles (80,000 kilometers) of rivers, and tens of thousands of acres of floodplains and estuaries.[16]

Lutz's work in the Connecticut basin is part of that vision. Her top priorities include restoring floodplains, making it easier for fish to migrate, and connecting streams throughout the basin. She drove me to the Knightville Dam on the Westfield River, a major tributary to the Connecticut in western Massachusetts, where the Corps and her team are studying the possibility of allowing small floods to pass through the dam in order to rejuvenate the river's floodplain. While Lutz's team has participated in the removal of Bartlett and 18 other dams in the basin, which has reconnected some 300 river miles, she also wants to see changes in how dams are operated. The Connecticut's large hydropower dams can make river levels fluctuate as much as 8 feet (2.4 meters) in 24 hours. Such big daily swings are completely unnatural to a river, and many fish and other organisms have difficulty coping with them. Five of the river's hydropower dams are now up for relicensing, an opportunity for federal regulators to set new operating requirements for the dam owners.

Better ways for fish to get past dams are also needed. Holyoke, the first big dam encountered when heading upstream on the Connecticut River, is equipped not only with fish ladders but also with an elevator to help American shad and other migrating fish pass the 30-foot-high granite structure in their search for spawning sites upstream. But of the roughly 400,000 shad that made it past Holyoke Dam in the spring of 2016, only 59,000 got beyond the next big dam at Turner's Falls, Massachusetts, Lutz said, and only 29,000 made it beyond Vernon Dam, just over the Vermont border.

Then there are the tens of thousands of structures called culverts around the basin that need upgrading and repair. Culverts are designed to channel streams under roads, but if built too small or with the wrong materials they can become major barriers to fish, turtles, salamanders, and other animals living in and along the river.

Lutz directed me to a culvert on Mitchell Brook in the town of Whately, Massachusetts. Before heading out, I examined a photo of the old culvert, an undersized piece of corrugated metal that was perched 2 feet above the stream's confluence with West Brook. Even a very athletic trout dying to reach its coldwater habitat could not have leapt into that culvert to continue its journey upstream. When I reached my marker, a pair of guardrails along the dirt road, I parked and walked through the woods to the stream. What I found in place of the old metal culvert was a smart-looking arch-shaped structure, wider than the stream channel.

The cobble substrate looked just like a stream bottom, and it touched the streambed on the other side—no large gap. While a driver crossing on the road above would notice little difference, for a trout on the move it was like night and day.

As stream temperatures rise with global warming, the native trout's ability to take refuge in cooler-water tributaries will become critical. Scientists with the US Geological Survey and the US Forest Service have been tagging and monitoring brook trout at this site since 1997 to better understand where they live and how they survive. The data gathered at the Mitchell Brook culvert is the only information of its kind and longevity in New England. With the upgraded culvert in place, more research will help reconnect river systems throughout the Northeast— and help fish, turtles, and salamanders negotiate the difficulties caused by roads crossing over streams.

At times the whole enterprise of river restoration can feel daunting. There are so many dams, culverts, and other barriers—physical, social, and economic—in the way. But then I stand beside a stream like Mitchell Brook and think that right here, in this place, scientists, conservationists, government agencies, and the community have come together to heal a stream and help a population of fish survive. They chose to act and make a difference.

~

The tension in the room was palpable. By June 2010, the worst drought in Australia's historical record had already lasted a dozen years. Now a scary dark cloud that threatened to unleash not rain but social disruption hung over the attractive harborside convention center in the New South Wales capital of Sydney. Hundreds of farmers, scientists, agricultural business representatives, and government officials had gathered to explore the future of irrigated agriculture in the Murray–Darling River basin, the Commonwealth's agricultural powerhouse. The authority that manages the basin's water was soon to release its findings on how much water needed to be cut from irrigators' allotments to make rivers and wetlands healthy again.

Much like the Colorado River in the US Southwest, the Murray–Darling basin is both southeastern Australia's lifeline and a vital asset to the nation's economy. The basin spans 14 percent of Australia's territory and supports 40 percent of its agricultural production. Farmers there grow cotton, rice, fruits, vegetables, alfalfa for dairy cows, and grapes for popular wines. It is also home to 30,000 unique wetlands, 16 of them internationally recognized, as well as a rich diversity of freshwater species, including the prized Murray cod.

What had farmers in a serious state of anxiety was a legal provision called a "sustainable diversion limit." Born of a national water act passed in 2007, the term refers to the volume of water that can be extracted from rivers and other water sources without damaging aquatic ecosystems or other key environmental assets. The nation of South Africa had made the environment a priority in a pioneering 1998 water act, but no other country had acted so boldly to reform water use to preserve ecosystems.

What struck me there in Sydney, though, was that almost everyone accepted the need for change, even though that change would bring hardship to some. The status quo—dried-up rivers, desiccated wetlands, and dying fish—was not acceptable. "We can do a thousand cuts and a generation of pain, or we can take our pill and move on," said Murray Smith, then CEO of the Northern Victoria Irrigation Renewal Project.

As the driest inhabited continent, Australia has always had its share of water challenges, and the basin that encompasses the nation's two longest rivers, the Darling and the Murray, is no exception. Some 94 percent of the watershed's annual precipitation evaporates or transpires back to the atmosphere. Just 6 percent replenishes rivers and groundwater, one of the smallest ratios in the world. To compensate for that arid-

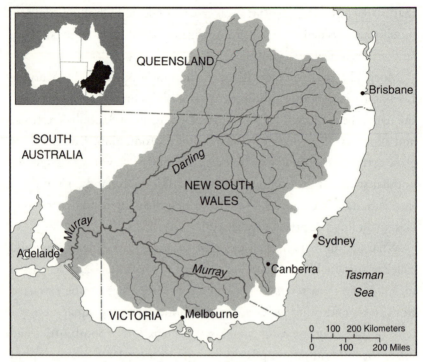

The Murray–Darling River basin in southeastern Australia.

ity and the wide swings in yearly precipitation and river flows, engineers built reservoirs able to store 130 percent of the basin's average annual runoff and a vast canal system to deliver that water to farmers with entitlements to it. But after river extractions more than tripled between 1944 and 1994, the Murray was in trouble. Severe low flows were occurring in 6 of every 10 years, compared with once in every 20 years prior to the dams and diversions.[17]

And then the unthinkable happened: the Big Dry.

For those of us who didn't live through it, the Millennium Drought is hard to fathom. It began in late 1996, according to Australia's Bureau of Meteorology, and grew in severity and duration to become the worst drought by far in Australia's record. Farm families that had worked their fields and milked their cows for decades watched their crops wither, sold their livestock, bulldozed orchard trees, and declared bankruptcy. Names kept getting added to suicide watch lists.[18]

Rivers, wetlands, and aquatic life fared no better. In 2002, the legendary Murray River ran dry for the first time. In the years that followed only dredging kept its mouth open to the sea. Beloved red gum trees fell sick and died along 900 miles (1,450 kilometers) of the river. Fisheries scientists were shocked when field surveys in 20 different river locations turned up no Murray cod. Black swan eggs, freshwater mussels, and other sacred totems of the Ngarrindjeri Aboriginal people disappeared.[19]

With policy as with technology, necessity is often the mother of invention. In the case of the Murray–Darling basin, innovative reforms had started in the midnineties after a toxic algal bloom spread through the lower third of the Darling River and the basin's water consumption climbed beyond sustainable levels. In 1995, the basin's Ministerial Council—a governing body consisting of ministers from each of the four basin states (New South Wales, Queensland, South Australia, and Victoria), the Australian Capital Territory, and the Commonwealth—placed an interim cap on extractions from the basin's rivers, and then in 1997 made the cap permanent.

In the water world, this was the first big experiment with cap-and-trade, a concept familiar by then in the realm of air quality. The cap on diversions meant that anyone who wanted water would have to purchase it from existing users, who, in turn, now had an incentive to be more efficient so they could sell whatever they saved. In short, water would be more productive and worth more money; there would be less waste, and the limited supplies would go to higher-value crops and other uses. A 1999 study projected that the cap and resulting trading activity would double the basin's economic value over 25 years.[20]

But there was a big problem. The Council had set the cap at a level of extractions that was already harming the Murray and its tributaries.

At best, the cap would stop further damage, not reverse it. And then the Big Dry kicked in, dropping river flows to record lows.

It was during this climate of drought and desperation that water management authority was wrested from the states and placed in a newly formed independent National Water Commission. The 2007 water act directed the Murray–Darling Basin Authority to develop a watershed-wide plan that would return enough water to rivers and wetlands to restore their health. To support the act's goals, the government committed more than A$10 billion to buy water entitlements from willing sellers and to invest in irrigation infrastructure and efficiency projects to reduce water waste. Scientists got busy assessing what level of extraction would leave enough water for ecosystems—the sustainable diversion limits that had the irrigators gathered in Sydney so worried.[21]

In October 2010, when the water authority released its findings in the *Guide to the Proposed Basin Plan*, it got hammered. The *Guide* called for basin-wide reductions in water diversions averaging 22–29 percent, with cuts in some areas as high as 35 percent. Across the basin, farm communities erupted in anger. Thousands turned out in the New South Wales town of Riverina to protest and heckle the water authority's chairman, Mike Taylor: "You're only worried about the basin. What about us?" one protester shouted, according to the Australian Broadcasting Corporation. "All I can see is my future and my children's future being flushed down the toilet," another cried out. One angry protestor threw the document into a bonfire. By day's end, the government announced a parliamentary inquiry into the human impacts of the Murray basin plan.[22]

It was a rude awakening for the water authority and an important lesson for many. Developing any policy or plan behind closed doors that affects livelihoods so deeply almost guarantees failure. For the Murray–Darling basin water authority, it meant a redo with a new approach: marrying sound science with real consultation with local communities.

In late November 2012, after a long period of comment on a new draft, environment minister Tony Burke unveiled the final plan. It is about a century overdue, Burke said, but "hopefully just in time" to save the river system. A month later he signed it into law.[23]

The plan calls for the return of 2,750 billion liters (726 billion gallons) to rivers and wetlands in the Murray–Darling basin, a volume scientists had determined would deliver significant benefits, but would fall well short of the amount needed to return the basin's ecosystems to health. Shortly before the final plan's release, the Australian government announced an additional A\$1.77 billion to secure an extra 450 billion liters (119 billion gallons) through efficiency measures on farms. A new agency, the Environmental Water Holder, would be in charge of delivering flows to designated ecosystems in accordance with the plan. By 2024, this environmental water bank would hold about 30 percent of all the water entitlements in the basin.[24]

For an overallocated, multistate river basin, the Murray–Darling plan was an extraordinary achievement. Scientists had prioritized where water could do the most ecological good. Many farmers had acquired faith in water markets. During the drought, the water available for irrigation dropped by two-thirds, but farm revenues fell by only one-fifth. Through more efficiency and trading, farmers found ways to boost water's productivity. Revenue per liter of water climbed two-and-a-half times. Back at the Sydney conference, Laurie Arthur, a rice farmer who served on the National Water Commission, told a packed room: "Trading is doing its job. It's helping irrigators adapt to changing circumstances."

A decade of shortage had also unleashed ingenuity and investment in smarter water management. The Coleambally Irrigation Cooperative in New South Wales, which is wholly owned by its farmer members, has automated its open-channel system and installed information technologies to enable precise control of water down to the level of its

491 irrigated farms. Automated flume gates control how much water is siphoned from the river into the canal network, while sensors in the soil monitor moisture levels, helping farmers know when and how much to irrigate. Other sensors detect leaks in the canals. Overall, the upgrades have reduced water losses by 70 percent. "That saved water goes back to our members as additional water," said John Culleton, Coleambally's chief executive. During the heart of the drought, when the cooperative got only 10 percent of its water allocation from the state, those water savings delivered an additional 17 percent, keeping some farmers in business, according to Culleton.[25]

But not long after I left Sydney, the rains returned. By September 2010, there was widespread flooding across southeastern Australia, devastating many communities. In an odd twist of fate, and perhaps a sign of the "new normal" of weather extremes to come, 2010 turned out to be the wettest year on record for the Murray–Darling basin. Heavy rains and floods continued into 2011. Memories of the Big Dry started to fade, and so did the urgency of returning water to the environment.[26]

"The Millennium Drought ended two years too early," said Keith Bristow, a hydrologist and soil physicist with CSIRO, Australia's national science agency, when I met up with him at a scientific conference in Phoenix, Arizona, in November 2016. I hadn't seen Bristow since the 2010 Sydney meeting, and his frustration with the recent turn of events in his home country was evident. "People just walked away from everything we'd put in place."

In 2014, the administration of Prime Minister Tony Abbott disbanded the independent National Water Commission, a "backward step," Bristow said. Oversight of water has shifted to the Commonwealth's Department of Agriculture and Water Resources, giving more influence to farmers. "Agriculture has an extremely powerful voice in the rural areas," Bristow said, and "farmers have really rebelled against some of the (environmental) targets that have been set." At this point, even the proposed 2,750 gigaliters of water, which scientists say is far too little for the environment, "may be hard to get." Indeed, just a couple weeks after I met with Bristow, the Murray–Darling Basin Authority recommended reducing water buybacks in the northern part of the basin where one town had seen employment fall by 21 percent since the diversion limits went into effect.[27]

For sure, Bristow said, the irrigation cutbacks have been hard on some rural towns. "It's tragic for those people, but in the scheme of things it's probably what has to happen. The drought forced us to do something. But the floods came too soon. We lost the momentum and the will."

The push–pull of water reform in the Murray–Darling will no doubt continue. Climate change loads the dice for more severe droughts and floods. The Aussies have shown they can act boldly—by saving water for the environment and establishing the best-functioning water markets in the world—to give their unique wetlands, beloved river red gums, and prized Murray cod a fighting chance. The Millennium Drought was a trial run for adapting to the new normal. It hasn't all stuck, but another test will almost certainly come with the next Big Dry.

~

A land of stately live oaks, swaying pines, and old cypress trees draped with Spanish moss, southwestern Georgia is a slice of rural beauty and Southern charm. Fields of peanuts, cotton, and corn stretch as far as the eye can see. But the lower Flint's $2 billion farm economy now faces a threat usually seen in the arid western United States rather than the humid Southeast: shrinking river flows and water disputes with neighbors.[28]

I traveled to the region in mid-August 2016 for a firsthand look at efforts to make agriculture more sustainable in the lower Flint River basin. With populations of rare freshwater mussels dwindling in the lower Flint, and Florida's leveling of a lawsuit against Georgia claiming overuse of shared waters, there's some urgency to the matter. Led by Casey Cox, who is equally comfortable talking about peanut farming, river conservation, and water technologies, the Flint River Partnership— a collaboration of the Flint River Soil and Water Conservation District, the US Department of Agriculture's Natural Resources Conservation Service, and The Nature Conservancy—believes smarter irrigation can be a big part of the solution.

The source of both the region's bounty and its trouble is an irrigation system called the center-pivot sprinkler. It consists of a giant horizontal sprinkler arm, often stretching the length of a football field, that pivots on wheels around a central point. While some systems rely on surface water, most in the lower Flint draw from the aquifer below. The irrigated field takes the shape of a giant circle, easily visible from the air. A typical center pivot might irrigate 200 acres (80 hectares).

In the 1970s, when center pivots rolled into southwestern Georgia, the farm economy took off. But so did water use. Peanuts and other field crops in the region typically require 22–25 inches (559–635 millimeters) of water. In a dry year, rain provides only about half that amount. To make up the difference, farmers activate their sprinklers and apply about a foot of water to their fields, or nearly 326,000 gallons per acre (3 million liters per hectare).

Today some 8,900 center pivots dot the landscape of the lower Flint. Farmers mainly draw water from an aquifer called the Upper Floridan, a highly productive karst formation that underlies most of Georgia's coastal plain. The aquifer is shallow, so drilling a well into it costs about a quarter as much as drilling down to a deeper formation. But the Upper Floridan is connected to the Flint River and its tributaries. Especially from June to October, when the stream is low, groundwater seeping into the channel can account for much of the total flow. As more water was pumped for irrigation, the stream levels during this critical period declined. The lowest streamflows in some tributaries dropped 30–40 percent. Long stretches of Spring Creek, historically a perennial stream, dried completely during eight summers between 2000 and 2011.[29]

The Flint River basin was once home to 29 species of freshwater mussels, one of the most diverse assemblages in southeastern North America. But the combination of drought and irrigation has caused their numbers to crash. As flows diminish, critical habitat along the banks disappears. Eventually, all that's left are shoals and then the streambed. Mussels, which aren't exactly fast movers, can get stranded, baking under a hot sun, with no oxygen, where they make a good snack for creatures passing by.[30]

While known by playful names like oval pigtoe, little spectaclecase, and shiny-ray pocketbook, mussels do serious work to help keep rivers clean and healthy. They filter pollutants from the water, move and cycle nutrients, and provide food for otters, muskrats, birds, and fish. "Mussels are a critical component of stream food webs and one piece of the equipment that makes a stream work well," said Stephen Golladay, an aquatic biologist at the J. W. Jones Ecological Research Center in Newton, Georgia. While Golladay emphasizes that there's a lot we don't know about mussels, his surveys find that populations are shrinking in the water-stressed middle reaches of Flint tributaries. "Spring Creek has seen pretty extensive die-offs," Golladay said.[31]

Today, the US Fish and Wildlife Service lists seven species of mussels in the Flint River basin as threatened or endangered, along with one species of fish, the Gulf sturgeon.

In 2000, in the midst of a severe three-year drought, the University of Georgia (UGA) opened a new research facility near the rural town of Camilla, Georgia. Called the C.M. Stripling Irrigation Research Park, its mission is to make irrigation more efficient in the southwestern part of the state. "The water wars were going on and the university realized it needed to do more in the lower Flint," said Calvin Perry, an agricultural engineer who runs the research center.

Perry was referring to the ongoing dispute between Alabama, Florida, and Georgia over the waters that make up the Apalachicola–Chattahoochee–Flint watershed, which extends from north of Atlanta south through Florida's panhandle to Apalachicola Bay. The Flint joins the Chattahoochee, which flows along the southern half of the Alabama–Georgia border, to form the Apalachicola, home to one of the highest concentrations of imperiled species in the United States. Apalachicola Bay ranks among North America's most productive estuaries, with highly prized harvests of blue crabs, shrimp, and more than 90 percent of Florida's commercial oysters. To reproduce, these fish require water that isn't too salty—which depends on freshwater flowing into the bay.[32]

The three states signed a compact in 1997 to develop a formula for dividing up the water fairly while protecting water quality, ecology, and biodiversity. Despite arduous negotiations, the compact expired in 2003 with no resolution. In October 2013, Florida sued Georgia in the US Supreme Court, effectively asking the court to limit Georgia's water use. While the "special master" appointed to oversee the case found in February 2017 that there is "little question that Florida has suffered harm from decreased flows," he recommended that the court deny Florida's request in part because the US Army Corps of Engineers, which controls flows through the basin's reservoirs, was not a party to the lawsuit, and he therefore could not devise a workable settlement.[33]

Meanwhile, Perry remained hard at work with his UGA colleagues and the Flint River Partnership to develop and spread more efficient ways of watering farm fields in the lower Flint. Many farmers have now added drop hoses with low-pressure sprays to their center-pivot sprinkler arms. Instead of spraying water high into the air, where wind and evaporation can rob up to 40 percent of the irrigation water, the upgraded sprinklers deliver bigger drops at lower pressures and closer to the crops, boosting efficiency from 60 percent to over 80 percent.

The team also saw great potential in better scheduling how much water got delivered to different parts of the farmers' fields. The result was an innovative technology called variable rate irrigation (VRI), which essentially tailors water application to field conditions. On average, about a tenth of each field in the lower Flint is taken up by roads, wildlife corridors, or wetlands, and is not growing a crop. VRI involves programming a GPS-equipped center pivot to shut off as it passes over those noncrop areas, meaning water use is cut by about 10 percent. Using a similar on–off mechanism, VRI also helps farmers apply less irrigation water where soils naturally retain more moisture. That typically brings water savings to 15 percent. If adopted widely in the lower Flint basin, these upgrades could mean significantly smaller water withdrawals from the Upper Floridan aquifer and area streams.

Despite its water-saving potential, VRI, which was commercialized in the mid-2000s, has been slow to penetrate the market. "Manufacturers will tell you it hasn't taken off yet," Perry said during my visit to the research park in August 2016. "Despite our many years of effort, it's still seen as an add-on."

VRI now qualifies as a standard conservation practice under US farm programs, which enables irrigators to get financial assistance to purchase it. But more incentives and technical assistance seem to be needed to expand its adoption. The cost of installing VRI varies with a number of factors, such as the brand and length of the center pivot, but runs about $14,000. Farmers can typically offset that investment through reduced pumping and fertilizer costs over the life of the system. But Cox, who works closely with the UGA team, feels farmers need more guidance with VRI systems. "The technology is ahead of the support capacity," she said. "It's all about helping farmers make better decisions."

With the lawsuit in process, the State of Georgia is hesitant to take much action. It did, however, reinstitute a moratorium on permits for new wells drilled into the Upper Floridan—essentially a cap on ground-

Casey Cox checks the control panel of a center-pivot sprinkler equipped for variable rate irrigation, which tailors water delivery to field conditions and can cut water use by up to 15 percent. Photo by Sandra Postel.

water use, albeit at levels already harmful to streams. George Vellidis, a UGA professor of crop and soil sciences, figures it would take $120 million to upgrade all center pivots in the lower Flint with standard VRI technology. "If the state said we're going to try to solve the water wars by investing in VRI, you'd have an explosion of use," he said.

Out in the field at the Stripling Research Park, Vellidis and Perry showed me the next generation of their technology, which they call dynamic VRI. It combines real-time soil moisture monitoring with VRI to schedule the timing and volume of irrigation much more precisely. The sensors, which are strategically placed in different parts of the field, give a numeric reading of soil moisture. That information is used to create a "prescription map," illustrating how much water to apply to each zone in the field on that particular day. An irrigator can send the map remotely to the pivot's VRI control panel.

In 2015, Vellidis and his colleagues worked with a local farmer to test the dynamic VRI process on a 230-acre (93-hectare) field of peanuts. Over the entire growing season, the dynamic VRI system recommended applying 30 percent less water to the field than did a commonly used irrigation program. The experimental and control fields got nearly the same yield of peanuts. Overall, dynamic VRI boosted water productivity—yield per unit water, or crop per drop—by an impressive 43 percent.[34]

But can dynamic VRI restore streamflows and help mussels and fish hang on in the lower Flint basin? The verdict is still out. More experimentation and research are needed. The people of Georgia and the lower Flint must decide if they want to take action to save the diversity of life in their streams. Perhaps future agreements or the courts will require that additional flow from the Flint must make it to the Florida border. But one way or another, as Vellidis put it, "it's got to be both a policy and a technology solution."

CHAPTER 11

Rescue Desert Rivers

This blue-green slice of life, fiercely bounded on every side,
will continue to try to persist for all it is worth.
—*Barbara Kingsolver*

PHILLIP DARRELL DUPPA, a late nineteenth-century pioneer of the
American Southwest, is a relative unknown in US history, but he war-
rants at least a historical footnote: he is credited with naming the capital
city of Arizona. With a classical education and five languages under his
belt, Duppa was no typical frontiersman. He drank and gambled, but
he also read Roman and Greek classics in their original language. Born
in France, he later traveled to America and eventually landed in central
Arizona. When it came time to anoint the new Anglo settlement that
had grown up along the Salt River, Duppa drew on his knowledge of
ancient mythology. The city should be called Phoenix, after the mythi-
cal sacred firebird that rises anew from its own ashes: "Prehistoric cities,
now in ruins, are all around you; a prehistoric civilization existed in this
valley. Let the new city arise from the ashes of those ruins."[1]

The civilization Duppa referenced was the Hohokam, a remarkable people who lived in the valleys of the Salt, Gila, Verde, Santa Cruz, and San Pedro Rivers of south-central Arizona for about a thousand years, roughly from 450 to 1450. The Hohokam formed one of the earliest and greatest irrigation societies of the Western world. In the lower Salt River Valley, the site of modern Phoenix, they built canal networks spanning 300 miles (480 kilometers), turning some 70,000 acres (28,300 hectares) of desert into flourishing fields of beans, corn, squash, and cotton. Villages sprang up about every 3 miles along the major canals, each covering about 15 square miles (40 square kilometers).

Like the much earlier and more famous irrigation-based civilizations along the Nile in Egypt and the Tigris–Euphrates in Mesopotamia, the Hohokam enjoyed food surpluses that freed up time for other endeavors. Many villages took pride in their ball court, a large oval depression where the community gathered for games and public events. From a Hohokam settlement called Snaketown, southeast of Phoenix, archeologists have unearthed beautiful beads, bracelets, pottery, and painted ceramics. At its zenith, Hohokam trade reached north to Chaco Canyon and south into Mesoamerica. The Hohokam population likely peaked between 50,000 and 200,000, and their territory spanned an area the size of Guatemala. With their cadres of skilled engineers, builders, artists, irrigators, farmers, athletes, entrepreneurs, and traders, they built a successful civilization that lasted a millennium.[2]

But then, sometime during the fifteenth or early sixteenth century, their society collapsed, and the Hohokam disappeared as a distinct culture. Scientists aren't sure what happened, but disruptive swings in climate are a leading contender. The downward spiral may have begun with a severe, two-decade drought in the American Southwest in the middle of the twelfth century. Connie A. Woodhouse at the University of Arizona in Tucson and her colleagues used paleoclimatic records to reconstruct annual Colorado River flows from 1146 to 1155 and found them to average 22 percent below those of the twentieth century.[3] While the Hohokam muddled through this "mega-drought," they were hit with another dry spell late in the thirteenth century and then a period of disastrous flooding early in the fourteenth. This wild, unpredictable weather may have created food shortages, hunger, and disease that in turn led to a cascade of falling population, labor shortages, decaying infrastructure, and social and political upheaval.[4]

However these factors played out, one emerges as key: an inability to adapt to a changing environment. In a way, the Hohokam's success may have been their undoing. Even with their extensive irrigation system, they could not successfully adjust when the climate shifted beyond the familiar range.

These challenges are eerily familiar to current residents of the American Southwest. As recently as 2000, Lake Mead—the vast reservoir on the Colorado River that supplies water to some 30 million people in Phoenix, Tucson, Las Vegas, Los Angeles, and other southwestern cities, as well as to several million acres of irrigated land—was nearly full. But by the spring of 2016, after a 15-year drought, it was down to 36 percent of capacity, the lowest level since it was first filled in the 1930s. Meanwhile, climate scientists expect that rising temperatures alone will cause the Colorado River's average flow to decrease by 5 to 35 percent by 2050, in large part due to higher rates of evaporation and transpiration. Should precipitation in the basin drop by even 5 percent, annual streamflow would decline by an additional 10–15 percent.[5]

All told, a future of "hot droughts" and less water in the American Southwest is likely if not certain. As always, there will be swings—floods following droughts, and cold spells in the midst of warming—but the overall trajectory is quite clear. Unless cities and farms find a way to cope with drier, less stable weather, they may find themselves in a situation like that of the Hohokam—with similarly apocalyptic consequences.

Facing these new realities will require the region to reform the policies and practices that have left it vulnerable. The binational treaty between the United States and Mexico known as Minute 319 (see Chapter 2) is a step in the right direction, as are the rules agreed to by the basin states in 2007 for sharing shortages. Another is a small pilot effort called the Colorado River System Conservation Program—a partnership of the US Bureau of Reclamation and the water authorities of Denver, Las Vegas, Los Angeles, and Phoenix—that is underwriting projects to lift the level of Lake Mead, a potential benefit to everyone in the basin. And proposals to decommission Glen Canyon Dam, once viewed as fringe, are now in mainstream discussions.[6]

But can southwestern rivers thrive in the new normal? More than 60 percent of the vertebrate animals of the American Southwest are "riparian obligates," meaning they can survive only in the gallery forests of cottonwoods, willows, and mesquite sustained by healthy rivers.[7] Three-quarters of the bird species that breed in the Southwest depend on these riverside corridors. Dams, diversions, and changes in land use have already wiped out as much as 90 percent of their habitat. How can we protect what remains, and, as in the Colorado Delta, bring back a portion of what has been lost?

An answer is emerging along three of the same desert rivers that sustained the thousand-year run of the Hohokam—the Gila, San Pedro, and Verde. There, farmers, ranchers, and conservationists are upending the conventional wisdom that water in the American West is all about fighting for the last drop. Instead, they are each taking some risks for the good of the whole—fish, wildlife, birds, beauty, livelihoods, recreation, and all else that benefits from a river flowing in the desert. One farm and ranch at a time, they are forging a future that might just keep the land, waters, and people ticking through the big changes ahead.

~

"I don't know what I pump and I don't care—and that's crazy," says Paul Schwennesen, a fit, energetic rancher in his late thirties. On his

The Gila River watershed in southwestern North America.

modest-size ranch, the Double Check, nestled in the lower San Pedro River Valley of southeastern Arizona, Schwennesen raises cattle to supply grass-fed beef to farmer's markets and restaurants in Phoenix and Tucson. About 1 mile (1.6 kilometers) of Schwennesen's land abuts the San Pedro. When he pumps groundwater to irrigate his fields, less seeps into the river channel. During irrigation season, the San Pedro shrinks to a trickle or completely dries up.

Unlike most irrigators, Schwennesen wants to be made to care how much groundwater he pumps. A decade ago, he took over operations at Double Check from his father, Eric, who now raises cattle in the high country near the Arizona–New Mexico border. Schwennesen, a graduate of the US Air Force Academy and Harvard University who deployed to a combat zone in Afghanistan, is a successful rancher and businessman. But he cares about the river too. In his mind, free water is no friend to the river, the economy, or the long-term health of the community, and he wants to see water better valued. "I am a free-market devotee," Schwennesen said. "Markets are the best way to allocate scarce resources. We'd love to see a market established for water."

Schwennesen is among a new cadre of farmers and ranchers who bring a more holistic, ecological way of thinking to land management.

"Water is the salient variable in these environments," he said, as we examined one of his experimental fields on a warm, late-May morning. "Anything you can do to alter the water regime is going to have the biggest effect. And the more organic matter we can squeeze back into the soil, the more water." That's a belief backed by science, and it's at the core of Schwennesen's mission. "Well managed land can give back more than it consumes," he said. "That's the miracle of it." (For more on regenerative farming and ranching, see Chapter 6.)

I've come to the Double Check Ranch smack in the middle of the driest time of the year, typically April to June. The much anticipated El Niño of 2015–16 did not deliver the rains most had hoped for. Just a short distance from where we stood, the San Pedro's channel was dry. Historically this portion of the lower river had flowed intermittently, but over time groundwater pumping and prolonged drought have depleted the base flows and dried up the channel.

The San Pedro is the last major undammed river in the American Southwest. Unlike the Colorado River and the Rio Grande, which flow south toward Mexico, the San Pedro originates in the state of Sonora, Mexico, and flows north some 160 miles (260 kilometers) before joining the westward flowing Gila River near the small copper-mining town of Winkelman. It is a winding ribbon of green through the desert that offers biological riches far out of proportion to its size. The forests that band both sides of the river provide one of the best remaining habitats for birds and wildlife in the Southwest. Millions of songbirds migrate along the San Pedro corridor as they journey between their wintering grounds in Central America and Mexico and their breeding grounds as far north as Canada.

By some estimates some 45 percent of the 900 bird species in North America make use of the San Pedro at some point in their lives. As habitat disappears along the Colorado and Rio Grande, more birds than ever seek out the San Pedro. The American Bird Conservancy calls the San Pedro the "largest and best example of riparian woodland remaining" in the Southwest. In addition to its rich diversity of birds, the river system is home to 180 species of butterflies, 87 species of mammals, and 68 species of reptiles. In the desert, animals naturally gravitate to water, shade, food, and protection, and the San Pedro is a rare provider of these necessities.[8]

But irrigated agriculture, copper mining, and, in the upper reaches of the valley, urban growth have put that wealth in jeopardy. And that's where Schwennesen comes in. He has partnered with the Tucson-based Arizona Land and Water Trust and hydrologists at the University of Arizona to see if he can maintain his profits while cutting water use by 20–30 percent. On experimental plots, he is shifting to less thirsty crops, mostly a mix of native perennial grasses and an annual rye crop. By direct-seeding the rye, he avoids the tillage that can erode and dehydrate the soil. With the deep root systems and active microbial community boosting organic matter, the soil will absorb and hold more water. Together, the new crops and healthier soil will mean less irrigation, and therefore more water for the aquifer that feeds the San Pedro.

As we hop a barbed-wire fence and head down to the river, bushwhacking through the dense forest, Scott Wilbor, then a project manager with the Arizona Land and Water Trust and an excellent birder, hears the chuckles, whistles, and clucks of a large warbler called a yellow-breasted chat. Soon after, Wilbor picks up the call of a gray hawk and a yellow-billed cuckoo. The birds are at home with the cottonwoods and willows that still grow along the San Pedro. Unlike in many riparian areas in the Southwest, the native trees have not yet been overrun by invasive tamarisk (also known as salt cedar).

As a young boy, Schwennesen had his mind opened to new ways of ranching and raising livestock when he lived with his family in Lesotho, a small kingdom encircled by South Africa. He tells the story of how his father, a range scientist, informed the local tribal people that they were overstocking their land with cattle by 300 percent. Not missing a beat, one of the villagers asked his father how that could possibly be true since they'd had abundant grass for hundreds of years. "The range model collapsed right there," Schwennesen said.

Today Eric and Paul practice rotational grazing, which involves keeping cattle bunched together and moving them frequently. The cattle's manure fertilizes the soil and their stomping breaks up the earth, allowing air and water to penetrate. The frequent moves prevent overgrazing and give grasses a chance to recover before the herd returns. The senior Schwennesen raises his cattle in the mountain grasslands near the Arizona–New Mexico border, and then brings them down to Double Check to add weight on irrigated pastures before slaughter.

But whether Double Check can thrive on substantially less irrigation water is the million-dollar question. The answer will affect not only the San Pedro and its bird life, but much of the West, where thirsty alfalfa and other types of hay top the list of water-guzzling crops. Under an arrangement with the Arizona Land and Water Trust, Schwennesen has agreed to cut his water use by 110 acre-feet (about 36 million gallons or 136 million liters) over two years on 24 acres (10 hectares) of experimental fields. That alone won't save the San Pedro, but if Schwennesen can make it work, others may get the confidence to shift to similar water-conserving measures.

Up until this partnership with Double Check, the Arizona Land and Water Trust had mostly freed up water for rivers by paying farmers to fallow their land. By contrast, this experiment keeps agricultural land productive while replenishing the aquifer and the river. The project is designed to pilot several different water-saving approaches and compare the results to determine which works best. Probes placed in the soil allow the team to calculate how much water the mix of native grasses and rye pasture consumes. A flow meter installed on Schwennesen's well measures his water use. For the 2016 irrigation season, the meter showed that Schwennesen had saved at least half of the two-year goal, putting him on track to achieve the planned reduction in groundwater pumping.[9]

"It's one of our most exciting projects because we've got all this monitoring going on," said Wilbor, "and Paul is coming from a conservation-minded background." As part of the agreement, the trust compensates Schwennesen for the value of the 110 acre-feet (135,680 cubic meters) he has agreed to leave in the aquifer, effectively putting a price on his water for the first time. That pleases Schwennesen, and, if they could weigh in, the willows and songbirds along the San Pedro would no doubt be pleased too.

~

In February 2012 as I was heading south from Phoenix, Arizona, I crossed a wide channel that was dry as a bone. Much to my dismay, it was the Gila River, the largest tributary within the lower Colorado basin. The sight was both sad and ironic in light of where I was headed—to visit Casa Grande, one of the famous ruins of the Hohokam. Much like the Colorado, the Gila (pronounced *HEE-luh*) is dammed and heavily diverted. Today only a small fraction of its historical annual flow of 1.37 million acre-feet (1.7 billion cubic meters) reaches its confluence with the Colorado near the US–Mexico border.

In many ways, though, the Gila is a tale of two rivers. In Arizona, the

Coolidge Dam turns its flow into San Carlos Lake, and canals siphon its water off to farms and cities. But upstream in New Mexico, conservationists proudly tout the Gila as the state's only undammed river: "Viva El Rio Gila—Wild and Free!" is a popular rallying cry. But with the river's freedom now in jeopardy from a proposed large-scale diversion, and with traditional irrigators in the valley depleting its dry-season flows, conservationists in New Mexico have stepped up efforts to both protect and restore it.

The Gila forms from springs and headwater streams in the high mountains of the Continental Divide of southwestern New Mexico. The Gila Wilderness, where three forks of the river converge, was the first federally designated wilderness area in the United States. After it coalesces, the Gila descends through the beautiful Cliff–Gila valley and then journeys westward toward the Arizona border.

With no dams blocking it, the upper Gila still flows to nature's rhythms. Its seasonal highs and lows and gentle meanders across a broad floodplain create a rich mosaic of habitats that supports a splendorous array of life. It is home to one of the most intact native fish communities in the lower Colorado River basin, including the threatened loach minnow, the spike dace, and the Gila trout. Like the San Pedro, its riparian habitat supports a spectacular diversity of birds, including the rare western yellow-billed cuckoo, the Gila woodpecker, Lucy's warbler, Bell's vireo, and perhaps the largest population of the endangered southwestern willow flycatcher, which fancies the Goodding willows that shade the Gila's banks.

Within minutes of my first visit to the river on a bright September day, I spied two kingfishers skimming the water. Then came a flicker of red—perhaps a vermilion flycatcher—and then what sounded, improbably, like a shrieking seagull, but what I learned was the call of a common black hawk, a stocky bird with a white-banded tail that lives in the riverside woodlands and preys on frogs and small fish. In the background was the music of the Gila's riffles, where the river bubbles over cobbles in its bed, adding oxygen to the water. It was a sensory delight, the sights and sounds of a living river.

In recent decades, threats to the upper Gila have surfaced a number of times. Conservationists have already blocked two proposed dams. But a new risk arose with passage of the federal Arizona Water Settlements Act (AWSA) of 2004, which authorized New Mexico to each year capture and use up to 14,000 acre-feet (17.3 million cubic meters) of the Gila's flow. Eager to exercise those rights, state officials proposed the construction of a large-scale diversion to meet the future water demands of southwestern New Mexico's farms and towns. But by skimming off the Gila's frequent flow pulses and small floods, the diversion would largely cut the river off from its floodplain, damaging the habitats that support the Gila's rich diversity.[10]

An initial appraisal by the US Bureau of Reclamation in 2014 found that none of the plans proposed for the diversion made sense economically: their estimated costs exceeded long-term benefits.[11] An analysis by Western Resource Advocates, a nonprofit organization based in Boulder, Colorado, found that local water conservation, recycling, and efficiency projects, along with already-planned local water transfers, could satisfy the region's water demands until 2050, with water to spare, and at a fraction of the cost of the proposed diversion.[12] Norman Gaume, an engineer and former director of New Mexico's Interstate Stream Commission (ISC), the agency pushing hard for the diversion, has called the options put forward by the ISC "technically flawed and financially infeasible."[13] And the nonprofit Gila Conservation Coalition, based in Silver City, New Mexico, which has led the charge against the diversion, advocated using available federal funds under the AWSA for cost-effective local projects rather than a "billion-dollar boondoggle" that could saddle New Mexico's taxpayers with a bill of hundreds of millions of dollars.[14]

Rationality has rarely prevailed, however, on questions of big water projects in the American West. As Daniel P. Beard, former commissioner of the US Bureau of Reclamation and author of *Deadbeat Dams*, remarked on a swing through New Mexico in November 2015, "We seem to have a need to build something—anything—even when the project makes no sense at all."[15] True to form, New Mexican officials informed the US Department of the Interior in late 2014 of their decision to move ahead with a diversion of the Gila. Ultimately, after economic and environmental reviews, including impacts on the upper Gila's seven threatened and endangered species, the US secretary of the interior will decide whether or not a diversion moves forward.

Meanwhile, in 2016, conservationists working to protect the Gila scored a quiet victory that helps bring New Mexico's water policies into the twenty-first century: water rights designated for irrigation were voluntarily used instead to replenish the river.

Although no dams block the upper Gila, earthen embankments divert water away from its channel and into ditches that supply irrigation water to local farms and ranches in the valley. As with the San Pedro, the diversions can leave long stretches of the Gila flowing dangerously low or not flowing at all for weeks at a time during the irrigation season. This poses problems not only for kayakers and river enthusiasts, but for fish and wildlife.

Martha Schumann Cooper, an ecologist with The Nature Conservancy (TNC), partnered with the Upper Gila Irrigation Association and the local community to try out a potential solution. TNC, which owns land and associated water rights in the Cliff–Gila Valley, and two conservation-minded neighbors who had chosen not to irrigate a portion of their land, volunteered to return some of their water to the river. In March 2016, the New Mexico Office of the State Engineer approved this use of their water rights, and that spring the Gila got its first "turn" on the ditch system. While this pilot project involved only a small volume of water, it set an important precedent for restoring river flows in New Mexico.[16]

~

The signature mantra of water law in the western United States—"use it or lose it"—is ingrained in just about every farmer's brain. The phrase stems from state laws that say if a water right is not put to beneficial use the state can take it back. The traditional thinking, said Kevin Hauser, a farmer who has lived in the Verde valley of central Arizona for nearly half a century, "is divert all you can and use all you can."[17]

So, many irrigators in the Verde valley were understandably suspicious when a hydrologist from Oregon named Kim Schonek arrived in 2008, tasked by TNC to protect the Verde River and its rich diversity of plants and animals. Rumors circulated around the community that TNC was out to get their water and put it back in the river.

Flowing 195 miles (314 kilometers) from spring-fed headwaters north of Prescott, Arizona, south to its confluence with the Salt River near Phoenix, the Verde joins the upper Gila and the San Pedro as the crown jewels of rivers in the Southwest. Even as it supplies the Arizona capital with drinking water, the Verde supports 92 species of mammals—including bobcat, gray fox, muskrat, and river otter—as well as populations of razorback sucker and other native fishes that are dwindling throughout the Colorado River system. It is also an avian paradise. Some 221 bird species nest, breed, or feed in the Verde's riparian forests of Fremont cottonwoods and Goodding willows. As on the upper Gila and San Pedro, these forests provide a critical flyway and nesting spot for the endangered southwestern willow flycatcher and numerous other migratory birds. Breeding bird densities in the Verde's riparian corridor are among the highest ever recorded in North America.[18]

Not surprisingly, TNC tagged the Verde as an "ecological hotspot"—

an environment rich in species diversity but at risk of losing much of it. Nearly one-third of the Verde's fish assemblage is federally listed as endangered or threatened.[19] In 2008 TNC scientists began a planning process to assess threats to the river. The group identified irrigation diversions as a major problem, Schonek said. "And no one was working on it."

Today, most farmers in the Verde valley irrigate the way their predecessors did in the 1860s. They build a simple earthen embankment in the river channel to divert flow into an irrigation ditch. Laterals off of the main ditches bring water to individual farms and properties. The system runs entirely by gravity. Seven main ditches, each managed separately by a board and "ditch boss," run through the valley. The ditch farthest downstream is the Diamond S, a 5-mile-long (8-kilometer-long) conveyor of water to landowners collectively irrigating about 400 acres (162 hectares) of crops and landscaping. During the summer irrigation season, the ditches would at times divert the Verde's entire flow. Most of the water not consumed by crops and vegetation would eventually make its way back to the river downstream, but for that several-mile stretch downstream of the Diamond S diversion, the Verde was severely depleted, if not completely dry.

Schonek knew her task would not be easy. Soon after she arrived in the valley, she began talking to the ditch boss, Frank Geminden, and other irrigators to learn how the ditch system worked. The fact that there was water at the end of the ditch meant that the Diamond S irrigators didn't need the amount they were taking out of the river. They took it because that was the easiest way to ensure that everyone on the ditch got water—and better to appear to use it than risk losing it. After absorbing the situation, Schonek designed a strategy that would benefit the river, the irrigators, and the broader community of Camp Verde, a town of some 11,000 residents located along the river.

Her solution was to provide incentives for the irrigators to divert less water, along with a technology—solar-powered automated headgates—that enabled them to do so. The idea was for the ditch system to deliver only what the users actually needed so as to leave more flow in the river. But first, Schonek had to build trust.

"I'm sure I met Kim over a beer," said Steve Goetting, a businessman, backyard pecan farmer, and chair of the Camp Verde Chamber of Commerce. In addition to long talks over local brews, Schonek organized a field trip. In December 2010, she took eight Verde valley irrigators, including Goetting, down to the Phoenix area to see some automated headgates in action. "I was completely enamored with them," Goetting recalled. She wanted to know, "What do we need to do to get these? How do we start?"

For Goetting, the river was more than a source of irrigation water for his pecan trees: it was the lifeblood of the local economy. He owned The Horn, a brewery and restaurant that served delicious pies made with his pecans. As chair of the chamber of commerce he viewed a healthier Verde River as a key to the valley's future. "The name of this town is Camp Verde," he said. "It would not do well as Camp Brown."

As Goetting's enthusiasm spread to the rest of the ditch board and then to the larger community, the initial discomfort with the idea of giving some water back to the river eased up. "This project sprang out of that field trip," Schonek said.

For the plan to succeed, it needed the right flow targets and incentives to meet them. Schonek and her TNC colleagues set a "minimum flow target"—the lowest the river should ever flow in the valley—of 30 cubic feet per second by 2020. That was about 43 percent of the historical low, but sufficient to meet key ecological and social goals. Schonek presented the Diamond S irrigators with a "diversion reduction agreement" that said TNC would pay for the automated headgate if they reduced their diversions by the agreed-upon amount.

The irrigators signed on in 2013. That year, they met the goal of leaving 5 cubic feet per second of flow in the river during the irrigation season, which runs from May 15 to September 15. In dry summer periods, that additional 5 cubic feet per second can nearly double the Verde's flow. TNC installed the automated headgate, which was ready to go for the next season. Once again, the irrigators met the flow target, and they used their financial reward to again make some upgrades. In this way, their water savings have grown year after year—and the river has gotten increasingly more flow. By 2016, the river received an additional 3,103 acre-feet (3.8 million cubic meters) during the irrigation season, more than double what it received in 2013.[20]

Schonek acted strategically in starting with the Diamond S, the most downstream diversion in the valley. Her plan is to gradually upgrade ditch systems one at a time, moving upstream through the valley. In this way, the water savings accumulate, and the extra flow remains in the river downstream through Camp Verde and beyond toward Phoenix. Just over 40 miles (65 kilometers) of the stretch below Camp Verde is officially designated a US Wild and Scenic River, one of the most beautiful and culturally significant stretches of river in Arizona.

Perhaps most importantly, no farmer or ditch member has sacrificed anything. "It's working very well," said Geminden, the ditch boss, who can operate the automated system from his cell phone. "We are still able to provide the water everyone needs."

A family farm run by Kevin Hauser and his son Zach is last in line for the ditch's water. The farm typically uses about half of the Diamond S water to irrigate their laser-leveled fields of alfalfa, vegetables, pumpkins, and sweet corn. The Hausers' satisfaction was a key barometer of the project's success. "If we're at the end of the ditch and we're satisfied, then everyone (on the ditch) should be satisfied," said the elder Hauser, who, as a young boy, spent many days down by the river. "I'm very happy with the in-stream flow agreement. The river offers a lot to the aesthetics of the valley."

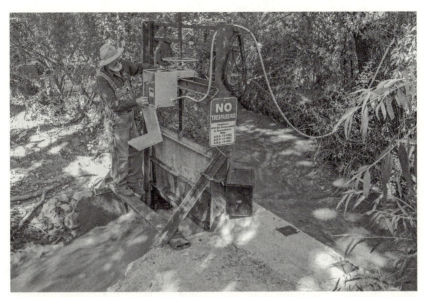

Irrigator Frank Geminden demonstrates how the automated headgate on the Diamond S ditch system keeps more water flowing in Arizona's Verde River. Photo by Cheryl Zook/National Geographic.

For Schonek, the success on the Diamond S inspired confidence to continue upgrading other ditches upstream. It has also brought in more funding, both from government sources and from corporations that see value in a secure water supply for the Phoenix area, a healthy river that provides recreational opportunities, and demonstrations of good water stewardship. For example, Recreational Equipment, Inc., better known as REI, is seeking to balance the water footprint of a new distribution center it is building in the greater Phoenix area that is certified for Leadership in Energy and Environmental Design (LEED). By investing in projects that restore flow to the Verde and its tributaries—including the conversion of inefficient flood irrigation to more efficient drip systems—REI is not only balancing its footprint, it is supporting outdoor recreational opportunities for its customers, its employees, and the general public.

TNC, the Friends of Verde River Greenway, and Change the Course, the national water restoration initiative I helped create, partnered to devise a portfolio of projects that benefited the river and also met REI's goals, which included volunteer opportunities for its employees. REI is also helping fund a conservation easement on a portion of the Hausers' farm to protect the river and riparian habitat from the impacts of future development. Altogether some half dozen corporations have partnered with Change the Course on projects to restore flow to the Verde and its tributaries, from ditch system upgrades to the installation of drip irrigation on valley farms.

In the summer of 2016, Friends of Verde River Greenway launched another creative mechanism to restore flow to the Verde: a voluntary water exchange. The Verde River Exchange connects residents and businesses in the valley willing to temporarily reduce their water use with others wanting to offset the impacts of theirs. In the first pilot, a local family agreed to forgo the irrigation of a small pasture for one year, generating water credits that will partially offset the use of groundwater by Merkin Vineyards and Page Springs Vineyards. (Groundwater sustains base flows in the Verde and its tributaries.) The two vineyards, in turn, will purchase the water credits, providing revenue that can be used to compensate the local family. It remains to be seen how much voluntary action the exchange generates. But it provides a new vehicle for those who want to do their part to sustain healthy rivers in their backyard.[21]

TNC is now scaling up its efforts in the Verde valley with a water fund similar to those it has created in the Rio Grande watershed and in a number of Latin American countries (see Chapter 3). The idea is to motivate downstream users—including drinking water providers and companies doing business in the Phoenix metropolitan area—to invest in projects in the valley that can improve water security. TNC aims to raise $20 million over the next decade to boost efficiency and upgrade infrastructure in the Verde valley, supporting healthy flows into the future.[22]

Arizona courts have yet to formalize rights to the Verde's water, and even though the ditch companies are some of the most senior users, the uncertainty is worrisome. But the partnerships in the Verde valley have increased water's value and created real benefits. They show that smart water use enables productive farming and healthy rivers to exist side by side.

CHAPTER 12

Share

One day you finally knew what you had to do, and began,
though the voices around you kept shouting their bad advice.
—*Mary Oliver*

ON DECEMBER 24, 1968, the crew of Apollo 8, the first manned mission to the moon, entered the lunar orbit. As Commander Frank Borman initiated a planned roll of the spacecraft, crewmember Bill Anders was photographing the moon's surface through a side window. All of a sudden, Anders caught sight of Earth emerging over the lunar horizon. With an "Oh my God" from Anders, Borman and James Lovell also took in the view of the exquisite blue-and-white orb. A calm but urgent search began for a roll of color film. "Hurry. Quick," Anders is heard uttering on the mission's tape recording. Lovell found the film just too late, but moments later, through a different window, the lonely planet appeared again. Anders then snapped what became one of the most iconic and transformative photographs of the twentieth century: Earthrise.[1]

Later that day, Christmas Eve, the astronauts held a live broadcast and beamed some of their pictures back to Earth. An estimated 1 billion people worldwide—one in four then living—viewed that broadcast. With little direction from NASA other than to say something appropriate, the astronauts read the story of Earth's creation from the book of Genesis. They closed with good wishes for "all of you on the good Earth."[2]

Neither the astronauts nor humankind would ever be the same. The Apollo crew had seen Earth, in Lovell's words, "as a grand oasis in the vastness of space." Anders later commented that they'd gone all that way to explore the Moon, but most importantly they had "discovered the Earth."

Earthrise showed our planet in its most fragile, isolated, and vulnerable state—aloft in outer space. Yet against the barren moonscape, the swirling blues and wispy whites turned Earth into a stunning gem in the vast darkness. Water, the source of life, and every astronaut's holy grail, cloaked the planet like a comfortable shawl. But most importantly, Earthrise showed humankind that it shares one finite, precious home.

It is no coincidence that the years immediately following the Apollo 8 mission witnessed an outpouring of activism and a host of new laws to protect the environment. Over the next five years, the US government established the Environmental Protection Agency and passed progressive laws to clean up the nation's air and water, to assess the environmental impacts of new development projects, and to safeguard other species from extinction. On April 22, 1970, some 20 million people turned out across the United States to celebrate the first Earth Day. Apollo 8's Earthrise joined Rachel Carson's *Silent Spring* and images of Ohio's oil-soaked Cuyahoga River catching on fire as catalysts for action.

But as we approach the 50th anniversary of Apollo 8, the mission to protect planet Earth and its inhabitants needs the equivalent of a booster rocket to thrust it into a new orbit. Massive change is literally in the air.

Atmospheric carbon dioxide levels have climbed above 400 parts per million, 14 percent higher than the 350 parts per million scientists say is the maximum safe level. Author Bill McKibben now calls our planetary home Eaarth to connote that it is no longer the same sphere that gave birth to and nourished human civilization. Even if we act urgently now to reduce emissions of greenhouse gases, as we unquestioningly should, our past emissions have locked the planet's climate and water cycle into decades of disruptions on a scale unprecedented in human existence.[3]

This is not a reason to descend into despair and wait for the dominoes to fall. Rather, in the vein of the farmers, ranchers, cities, communities, scientists, and conservationists profiled in this book, it is time to boldly adapt how we live, work, and manage our lands and watersheds to prepare for what's coming—more extreme floods, droughts, and fires; the drying of soils and streams; the shrinking of lakes, the contraction of wetlands, the loss of habitats, the expansion of dead zones, and the decline of fisheries and aquatic life. Our challenge as a society is to build resilience—the ability to cope with disturbance while continuing to function. As the stories in this book have shown, replenishing the world's natural flow of water is among the best ways to build that resilience.

To succeed, however, we must first come to grips with the primal revelation of Earthrise. The beautiful blue sphere the Apollo 8 astronauts beamed back to Earthlings on that December night a half century ago showed us unmistakably that water is the planet's greatest gift, the gift of life—and that it is finite. From that knowledge flows a moral truth: if water is essential to life, and it is finite, the ethical response is to share it with all of life.

~

By late spring 2012, as drought spread across much of the United States, conditions looked grim for the Yampa River in the upper Colorado

River basin. The Yampa flows 250 miles (400 kilometers) across a storied valley of farms and ranches in northwestern Colorado, through the popular tourist town of Steamboat Springs, and then west to Dinosaur National Monument, where it joins the Green River near the Utah border. On May 1, the snowpack that feeds the Yampa stood at only 17 percent of the long-term average. By June 27, the river's flow had dropped to just 5 percent of normal for that time of year. Locals worried about a reprise of 2002, the last serious drought in the region, when extreme low flows killed fish deprived of oxygen or food, and forced the shutdown of tubing and fly-fishing businesses at the height of the summer tourist season.[4]

"It looked like a system that was ecologically going to crash," said Amy Beatie, executive director of the Denver-based Colorado Water Trust. "The river was starting to crater." At risk along with local business revenues was the population of native mountain whitefish, a close cousin of salmon and trout that is found in only two river basins in Colorado. The whitefish had taken a bad hit during the 2002 drought.

But then on Friday, June 29, as if by a miracle, the river started to rise. By later that night the Yampa's flow was nearly out of the danger zone. Something unexpected had happened: an intervention to spare a river and its dependents from decimation during a drought.

Back in the spring, when the paltry mountain snowpack spelled disaster for so many of Colorado's rivers and streams, the Colorado Water Trust had issued a statewide request for water. Anyone willing to sell or temporarily lease water was encouraged to contact the trust. One responder to the call was Kevin McBride, director of the Upper Yampa Water Conservancy District in Steamboat Springs. McBride had just had a contract with a customer fall through, leaving 4,000 acre-feet (4.9 million cubic meters) of Yampa River water unclaimed in Stagecoach Reservoir.

"We rocketed that [project] to the top of our priorities," said Beatie.

For a total of $140,000, a very reasonable $35 per acre-foot, the trust leased the district's spare water and had it released gradually from the upstream reservoir from late June to September so as to keep the Yampa as healthy as possible throughout the summer. The leased water generated extra hydropower at Stagecoach Reservoir, helped raise oxygen levels for fish, and allowed sports like tubing and fly-fishing to resume— saving businesses in Steamboat Springs tens of thousands of dollars. The co-benefits made this project "a win-win-win-win," Beatie told me shortly after my visit to the Yampa Valley that summer.

Besides rescuing a river, the life within it, and the local economy, the Yampa drought lease set a precedent in Colorado. It was the first use of a 2003 state law, enacted in the aftermath of the 2002 drought, that allows farmers, ranchers, water districts, and other entities to temporarily loan water to rivers and streams in times of need. Colorado, like most western states, has historically required water rights to be used for specified "beneficial" purposes, such as irrigation or drinking water supplies. But now the Colorado legislature was saying something new: under certain conditions, it was possible to share water with a river without risking the loss of one's water right.

Armed with that new tool, the Colorado Water Trust executed another lease for the Yampa the following year, and has gone on to construct deals to boost flows in the Fraser and other Colorado rivers and streams. The Yampa story shows that the combination of sensible laws and willing partners can help heal a river. These are acts of stewardship—of sharing the gift of water with the ecosystems that sustain life on Earth.

Such acts are occurring all around us and take many forms. Shortly after I returned from the Yampa valley in August 2012, I learned that the third-largest river in New Zealand, the Whanganui, had been granted a voice under the law, similar to that granted to a person or a company. In one of New Zealand's longest running court cases, the local Māori people, the Whanganui River iwi, had won for the river the status of an integrated living whole, *Te Awa Tupua*, with rights and interests. Two guardians, one appointed by the iwi and the other by the Crown, will protect those interests.[5]

The New Zealand agreement came four years after a new constitution in Ecuador granted legal rights to rivers, forests, and other natural entities. And in 2011, the high-elevation Andean nation of Bolivia passed laws granting nature equal rights to humans, building on indigenous beliefs that the earth deity known as Pachamama is the center of all life.[6]

The granting of legal rights to ecosystems might sound odd and extreme. But is it more radical than depriving a river of water, the very thing that makes it a river? Even if we do not grant nature the rights of a person, we can still bring it into the prevailing legal system.

In essence, legal scholar Christopher D. Stone made that argument in his famous 1972 essay "Should Trees Have Standing?" Stone maintained that rivers, trees, and other "objects" of nature do have rights, and that those rights should be protected by granting legal standing to guardians of those "voiceless entities of nature." These ideas resonated with US Supreme Court Justice William O. Douglas, who that same year wrote a dissenting opinion in the case of *Sierra Club v. Morton*, in which he argued that natural entities should have standing so that legitimate legal claims could be made for their preservation. The river, Douglas wrote, "is the living symbol of all the life it sustains or nourishes—fish, aquatic insects, water ouzels, otter, fisher, deer, elk, bear, and all other animals, including man, who are dependent on it or who enjoy it for its sight, its sound, or its life. The river as plaintiff speaks for the ecological unit of life that is part of it."[7]

Stewardship of Earth's finite freshwater is all about finding an entry point and then taking action. For some, that means bringing about change through the channels of policy or law. For many conservationists, it is taking on strategic projects that return water to a damaged ecosystem. For a growing number of businesses, it is about reducing the environmental effects of their products. And for others, it is acting locally to better a river or watershed they know and love.

Such was the case for Roger Muggli, a third-generation farmer in the Yellowstone River basin of eastern Montana. I visited him at his farm outside of Miles City on a warm summer day in August 2013. Muggli's call to repair the Tongue River, a major tributary to the Yellowstone, came at the ripe age of 10 when he was shocked to find fish flopping around on his family's farmland. A diversion structure, 12 Mile Dam, was blocking saugers, suckers, and sturgeon from reaching their spawning grounds. Aside from that, some fish floating downstream were getting swept into the local irrigation canal and spewed onto his family's cropland.[8]

"I couldn't stand to watch them die in the field," Muggli recalled. He would scoop stranded fish into buckets, hop on his bike, and release them into the Yellowstone River, which flowed near the Muggli farm. He tried placing a screen against the canal's headgate, but it got so clogged with debris it would no longer pass irrigation water. The week he got his driver's license he took a bucket of fish from his field down to local officials, convinced that they would be mortified like he was at the death of so many innocent creatures. Instead they shooed him out of the office. Undeterred, and fortified by his mother's reminder that he would outlive the older folks shouting him down, he waited for his opportunity.

It came in 1986, when, in his late thirties, Muggli was elected to the board of the Tongue and Yellowstone (T&Y) Irrigation District. Both his father and his grandfather had served before him, so the Muggli name was well known around Miles City, where he'd lived his whole life. In addition to farming 1,700 acres (690 hectares), Muggli owns and operates the largest feed-making plant in Montana. In the off-season, he and his team turn alfalfa and grains into some 30,000 tons of pellets that feed livestock throughout the state.

From his position on the T&Y board, Muggli went about building a coalition of partners to solve the problem that had plagued him since he was a boy. The first fruits of his crusade materialized in the late 1990s, when, with support from public agencies and private conservation groups, he oversaw the installation of a new headgate system on the canal. Muggli showed me an inlet with a baffled wall that allows water through but keeps fish out, and a passageway that guides fish carried into the inlet right back to the river on the dam's downstream side. As a result, many thousands of fish previously siphoned into the irrigation canal to meet certain death were now heading downstream to their mother river, the Yellowstone, which ultimately joins the Missouri, the nation's longest river.

"You've got to fight it to the end or it's not going to get fixed," Muggli said.

In 2007, Muggli's full dream for a fish-friendly diversion dam was realized. With assistance from state and federal agencies, as well as The Nature Conservancy, a $400,000 bypass channel was built to enable fish heading upstream to spawn to circumvent 12 Mile Dam. State fisheries assessments have found that the Muggli Bypass is helping a wide variety of native fish get past the dam. Shortly after the upgrade at 12 Mile, two other diversion structures farther upstream on the Tongue began to be dismantled. Combined with the Muggli Bypass, their removal opens up some 190 miles (306 kilometers) of river habitat and spawning grounds on this critical tributary. Success on the Tongue has also inspired fish-passage projects on the Yellowstone, in particular to aid the ancient and endangered pallid sturgeon.[9]

Ambitious efforts like these require a great deal of collaboration and commitment among many agencies and stakeholders. But the spark for stewardship came from a 10-year-old boy's resolve to do something about the stranded and dying fish in his family's fields.

"It's a complex world and we have to be responsible players in it," Muggli said. "I want to leave this place better than I found it." And as for those naysayers that his mother advised him about: he outlived them all, he said. "They're all pushing up daisies."

~

Back in the late spring of 1992, after I'd finished drafting *Last Oasis*, my first book on global water issues, I set it aside for a few days. Something didn't feel quite right. I had written all 13 chapters I'd outlined; I thought I was done. But as I gazed out the upstairs window of my little rental home in the northwestern corner of Washington, DC, I knew something was missing. And then thoughts came to mind of the conservationist Aldo Leopold and his exposition of the land ethic. Leopold maintained that an extension of ethics to our relationship with the land and to the animals and plants that grow upon it was "an evolutionary possibility and an ecological necessity." That was the piece I was missing.

Laws, regulations, and markets were essential tools to protect water's place and functions in the natural world, but they weren't sufficient. We needed a water ethic to guide us to a new way of relating to freshwater. "We have been quick to assume rights to use water," I would go on to write, "but slow to recognize obligations to preserve and protect it." Living by a water ethic means "using less whenever we can, and sharing what we have."[10]

For the last 25 years, that idea of using less and sharing more, especially with the natural world, has framed my work. Without dedicated action, an ethic is an empty platitude. Earthrise-inspired stewardship—the recognition that all of life on this blue orb is connected by water and we must therefore share water with all of life—would take action by all segments of society. And so it was on a spring day in 2011 that I picked up the phone in my New Mexico office and embarked on an inspired conversation with an entrepreneurial freshwater biologist named Todd Reeve, whom I'd never met.

At the time, I was serving as lead water expert for a new freshwater program at the National Geographic Society. I'd worked with a creative team to educate and engage the public on freshwater issues. We built a calculator online that enables users to determine their personal water footprint—the volume of water it takes to sustain their diet and their energy use, and to generally keep their daily lives afloat—and we offered tips on how to shrink that footprint. The calculator was getting used. Awareness was building. *Time* magazine had given a shout-out to our website, and *New York Times* columnist Nicholas Kristof had directed readers to our water footprint tool. But what about the sharing part? How do we motivate the return of water to depleted rivers, shrunken wetlands, and the birds and wildlife that depend on them?

That was the question behind my phone call to Reeve, then vice president and now CEO of the Bonneville Environmental Foundation (BEF), a small nonprofit organization based in Portland, Oregon, dedicated to renewable energy, climate change mitigation, and watershed protection. I had just learned about Reeve's Water Restoration Certificate, a program that helps businesses balance their water footprints by investing in projects to restore water to depleted ecosystems. During an inspired brainstorm, we sketched out a water stewardship initiative that would bring the public, businesses, and conservation groups together to do the two big things necessary for healthy economies to exist side by side with healthy rivers: shrink our human water footprint and restore water to the natural world. We would make everyday people our partners, promising that for every pledge to conserve, we would return 1,000 gallons of water to a depleted ecosystem. Companies would underwrite those pledges and balance their own water footprints by funding on-the-ground restoration projects. We would identify and support scientifically sound and ecologically beneficial projects that restored freshwater ecosystems. The goal was to create a virtuous and expanding cycle of freshwater education, conservation, and restoration.

The initiative came to be called Change the Course, and its founding partners were BEF, the National Geographic Society, and the social-action enterprise Participant Media. For three years, we piloted Change the Course in the iconic Colorado River basin, from headwaters in the Rocky Mountains to the delta in Mexico. The Yampa drought lease I described earlier in this chapter was one of our kickoff projects, and it showcased the kinds of efforts we most wanted to support—those that demonstrate how innovative policies, technologies, and practices can restore rivers, aquifers, and wetlands while also benefiting local communities and economies.

Change the Course supported a number of restoration efforts profiled in this book, including projects in the Colorado River Delta in Mexico, as well as the San Pedro, the Gila, and the Verde. In more recent years, we have expanded our work to California, Georgia, and other parts of the United States. As of early 2017, nearly a quarter of a million people have joined our conservation community, 30 diverse companies have signed on as sponsors, and we have supported 33 on-the-ground projects, from aquifer recharge and river flow enhancement to floodplain protection and the creation of wetland habitats.

The billions of gallons we have restored to the environment are but a drop in the bucket of what is needed. But it's the start of a movement to inspire every individual, business, and organization to do what it can to conserve, and then go beyond and give some water back to nature. A growing number of food and beverage companies, for example, now recognize that, when the water consumed by crops in the field is taken into account, their products are among the most water intensive to produce. Good stewardship means balancing those impacts. As

Deanna Bratter, director of corporate sustainability at Colorado-based WhiteWave Foods writes in the *Huffington Post*: "Companies need to go beyond facility footprint, taking responsibility for the water needs of our communities, supply chains, and the world at large."[11]

Sports teams are taking on water stewardship, as well. To remind its fans that ice rinks depend on water, the National Hockey League runs a "Gallons for Goals" program that restores 1,000 gallons to an ecosystem in need for every goal scored. To date, the NHL has restored 50 million gallons (189 million liters) through projects that meet BEF's restoration criteria. The New York Mets baseball team is stepping up to the plate as well. In 2017 the Mets will work with Change the Course to balance the water footprint of the team's home stadium, Citi Field, for the season by supporting flow restoration projects.

Stewardship is more than a feel-good concept; it must yield results. The proof is in the pudding. On balance, are rivers getting healthier, aquifers being recharged, floodplains being rejuvenated, and wetlands being expanded? Are we becoming more resilient to droughts, floods, and fire? Is the water cycle being replenished and repaired?

So far the answer is no. At best, it's one step forward, two steps back. But here and there, we're making progress. There's inspiration in transformations like that in Arizona's Verde Valley, one of the few remaining healthy riparian areas in the American Southwest, where conservationists, farmers, local businesses, and some half dozen corporations have come together to return water to the Verde and enable the river to flow continuously again. There's inspiration, as well, in Australia's Murray–Darling basin, where government agencies have adopted policies to rebalance water use in nature's favor. And also in Mexico's Colorado River Delta, where two nations worked creatively with scientists and conservationists to give water back to an ecosystem once written off as dead. The delta's revival speaks not only to the resilience of ecosystems but also to what we humans can accomplish when we put our minds, hearts, and hands to the task.

We can choose to write a new water story. Depletion and dead zones are not inevitable. Even as dams go up, others are coming down, allowing fish to reach their spawning areas and opening up vast stretches of habitat. Even as aquifers are depleted, progressive farmers and cities are banking water underground to prepare for the dry days ahead. Even as populations grow, new technologies for treatment, recycling, and reuse are turning wastewater from a costly nuisance into a valuable new supply. Even as forest fires burn hotter and wider, coalitions of conservationists, businesses, and government agencies are rehabilitating watersheds to safeguard drinking water and communities downstream. Even as food demands rise, the marriage of information technologies with efficiency technologies is helping farmers get more crop per drop, saving water for rivers and aquifers. And even as land is degraded, ranch managers are employing cattle to revive grasslands, sequester carbon, improve soil moisture, and recharge groundwater. Yes, the water cycle is broken, but one river, one wetland, one city, one farm at a time, we can begin to fix it.

If the twentieth century was the age of dams, diversions, and depletion, the twenty-first century can be the age of replenishment, the time when we apply our ingenuity to living in balance with nature. In so doing, we can quench our own thirst while leaving a healthy water cycle for future generations. As we shift from the utilitarian view of water as a "right" and a "resource" toward the Earthrise view of water as the planet's greatest gift, our moral compass will direct us toward sharing water, not only among ourselves but with all living things.

In the end we will discover that what seem like acts of altruism or stewardship in fact serve our own interests—because water connects us to all of life. We know this in our bones.

Acknowledgments

I'VE LEARNED NEVER TO DOUBT that a phone call can change your life. A 2009 call from Barbara Rehm, one of the most creative and inspiring people I've ever known, changed mine. It led to a six-year fellowship with the National Geographic Society that resulted not only in this book but in a childhood dream coming true. To Barbara, Alex Moen, Cheryl Zook, David Braun, and our whole amazing Nat Geo team, thank you for the opportunity and support you provided to advance freshwater conservation and launch Change the Course.

I am forever indebted to Todd Reeve for not putting the phone down on that spring day in 2011 until we'd brainstormed our way to Change the Course. Our shared vision and passion to restore the natural world inspired me every day while writing this book. Todd, Val Fishman, and Sara Hoversten at the Bonneville Environmental Foundation have taught me that a small group of people—along with great partners—can change the world.

I'm grateful to the Walton Family Foundation for supporting my work while I was with National Geographic, as well as my travels to tell the stories of water restoration in the Colorado River basin. Special thanks as well to the C. S. Mott Foundation for a generous grant to support the research and writing of this book.

Many people shared their time, insights, and experiences with me, whether through formal interviews, casual discussions, and email exchanges, or during long days out in the field. I am grateful to Beth Bardwell, Amy Beatie, Cameron Becker, Nick Blom, Margaret Bowman, Keith Bristow, Christina Buck, Juan Butrón, Yamilett Carrillo, Martha Schumann Cooper, Casey Cox, Peter Culp, Helen Dahlke, Mary Ann Dickinson, Andrew Fahlund, Jay Famiglietti, Karl Flessa, Kirk Gadzia, Tamara Gadzia, Edwina von Gal, Jennifer Garvey, Frank Geminden, Edward Glenn, Christopher Gobler, Steve Goetting, Stephen Golladay, Brian Graber, David Groenfeldt, Cheng Guangwei, Jerry Hatfield, Kevin Hauser, Zach Hauser, Charles Hofer, Osvel Hinojosa Huerta, Zhang Junzuo, Nancy Kelley, Eloise Kendy, Estella Leopold, Gabriele Ludwig, Kim Lutz, Kevin McBride, Laura McCarthy, Daniel Mountjoy, Roger Muggli, Adrian Oglesby, Sam Passmore, Calvin Perry, Liz Petterson, Jennifer Pitt, LeRoy Poff, Wesley Porter, Robert Potts, Jon Radtke, Nancy Ranney, Todd Reeve, Brian Richter, Jeffrey Romanowski, Lauret Savoy, Karen Schlatter, Kim Schonek, Judith Schwartz, Paul Schwennesen, Laurie Seeman, Amy Singler, Allyson Siwik, Virginia Smith, Morgan Snyder, Phoebe Suina, George Vellidis, Amy Vickers, Casey Wade, Andy Warner, Scott Wilbor, Kate Williams, Sharon Wirth, Gary Wockner, Rebecca Wodder, Priscilla Solis Ybarra, Katherine Yuhas, and Francisco Zamora.

A number of friends and colleagues I just named also took time out of busy schedules to review drafts of various chapters, for which I'm very grateful. Any errors that remain reside at my doorstep.

A very special thanks to Brian Richter who for the last 17 years has been the best sounding board, collaborator, and coauthor one could hope for. His contributions to river restoration run deep and wide.

Revisiting my notes from my 1988 trip to China's Loess Plateau led me to search for Junzuo Zhang to express my gratitude for all she did to make that trip so memorable. Reconnecting with Junzuo was an unexpected highlight of *Replenish*.

I must give a shout-out to Circle of Blue, a small band of dedicated journalists that does a remarkable job reporting on the world of water and keeping us water wonks up to date. Thanks to the Water Footprint Network, as well, whose work and numbers I have relied upon for years.

It takes a village to produce a book, and I cannot thank Island Press enough for partnering with me once again. Emily Turner is a writer's dream editor. She helped me frame what I wanted to say into a coherent narrative, and when it came time to edit my words, gracefully simplified my message without once changing my meaning. Many thanks, too, to Diane Ersepke for her expert copyediting and to Sharis Simonian for shepherding *Replenish* through production. Chris Robinson prepared all the maps, and right from his first draft I knew this task was in the right hands. I am also grateful to Cheryl Zook at National Geographic for the use of photos she took while we were in the field, as well as for the use of photos provided by Craig D. Allen at the US Geological Survey, Katherine Kerlin at UC–Davis, Terrie Wade from Marfa, Texas, and Katherine Yuhas at the Albuquerque–Bernalillo County Water Utility Authority.

As I was writing *Replenish*, the death of my father, Harold, to whom I owe the moon, and the births of sweet Azalia and Abigail, reminded me how life, as water, cycles and cycles, and how precious it is. Thanks to my wonderful family—for everything.

Lastly, I am grateful beyond measure to Virginia Smith, my best friend, loving partner, and wisest critic. Her deep belief that we can succeed against the odds in making this world a better place—a belief she lives out every day—gave me the fortitude to anchor this book in realistic optimism. I dedicate *Replenish* to her.

Notes

Note: The following notes are presented in an abbreviated fashion. Full citations for each publicly available source can be found in the Bibliography.

Chapter 1

1. City of Fort Collins, Colorado, "2013 Flood"; quote from Kimsey, "Rainstorm of 'Biblical Proportions.'"
2. US National Oceanic and Atmospheric Administration (NOAA), National Weather Service, "The Record Front Range and Eastern Colorado Floods."
3. Howard, "Amid Drought, Explaining Colorado's Extreme Floods."
4. Gillis, "Climate Chaos, Across the Map"; Kozacek, "Ethiopia Hunger Reaches Emergency Levels"; Davies, "120,000 Still Displaced by Flooding Rivers"; British Broadcasting Corporation, "Flood Bill for Winter Estimated at 1.3bn"; Feaster et al., "Preliminary Peak Stage and Streamflow Data"; United Nations Children's Fund (UNICEF), "Malnutrition Mounts as El Niño Takes Hold."
5. US National Oceanic and Atmospheric Administration (NOAA), "2016 Marks Three Consecutive Years of Record Warmth for the Globe"; twentieth consecutive year from NOAA, National Centers for Environmental Information, "National Overview—Annual 2016."
6. United Nations, Office for Disaster Risk Reduction and the Centre for Research on the Epidemiology of Disasters, "The Human Cost of Weather Related Disasters 1995–2015," 2015.
7. Lesk et al., "Influence of Extreme Weather Disasters on Global Crop Production."

8. Hower, "Report: 91% of Top Firms View Climate Shocks as Business Risk."

9. World Economic Forum, "Global Risks Report 2016."

10. Kandel, *Water from Heaven.*

11. One cubic kilometer is the volume of a cube 1 kilometer long, wide, and deep. It equals 1 billion cubic meters or 1 trillion liters. In standard US usage, the equivalent is 264 billion gallons.

12. The components of the global water cycle are estimates and vary to some degree by source. I used the approximations published in Jackson et al., "Water in a Changing World."

13. UN Food and Agriculture Organization, Aquastat Database; Population Reference Bureau (PRB), "2015 World Population Data Sheet."

14. Postel, *Pillar of Sand: Can the Irrigation Miracle Last?*, 13–28.

15. Jacobsen and Adams, "Salt and Silt in Ancient Mesopotamian Agriculture"; Lagash and Umma from Postel and Wolf, "Dehydrating Conflict."

16. Han dynasty dams and 1885 start of Indus scheme from McNeill, *Something New Under the Sun: An Environmental History of the Twentieth-Century World,* 157–60; salt buildup and irrigation area from Postel, *Pillar of Sand,* 58, 96.

17. Hays quoted in Pisani, "Water Planning in the Progressive Era."

18. McNeill, *Something New Under the Sun: An Environmental History of the Twentieth-Century World,* 149.

19. McGee, "Water as a Resource."

20. Nehru quote from Radkau, *Nature and Power: A Global History of the Environment,* 176; D'Souza, "Framing India's Hydraulic Crisis: The Politics of the Modern Large Dam."

21. Water Footprint Network (WFN), online product gallery at waterfootprint.org. WFN is based in the Netherlands.

22. National Geographic Society, Water Footprint Calculator, Methodology and Tips. Our team used the best data we could find for each major water-footprint category, but the results are only approximations of one's actual water footprint.

23. "2015 World Population Data Sheet."

24. Hoekstra and Mekonnen, "The Water Footprint of Humanity"; Jaramillo and Destouni (2015) report a trend of increasing relative evapotranspiration from reservoirs and irrigated agriculture that suggests humanity's

annual water footprint may be closer to 10,688 billion cubic meters (2,823 trillion gallons).

25. Interception of river flows from Vörösmarty and Sahagian, "Anthropogenic Disturbance of the Terrestrial Water Cycle"; 100 billion tons from Syvitski et al., "Impact of Humans on the Flux of Terrestrial Sediment to the Global Coastal Ocean."

26. Richter et al., "Lost in Development's Shadow: The Downstream Human Consequences of Dams."

27. Ricciardi and Rasmussen, "Extinction Rates of North American Freshwater Fauna."

28. Winemiller et al., "Balancing Hydropower and Biodiversity in the Amazon, Congo and Mekong."

29. Diaz and Rosenberg, "Spreading Dead Zones and Consequences for Marine Ecosystems."

30. Groundwater from Postel, *Pillar of Sand*, 80.

31. Pomeranz, "The Great Himalayan Watershed: Agrarian Crisis, Mega-Dams and the Environment"; Jha, "India's River-Linking Project Mired in Cost Squabbles and Politics."

32. Winemiller et al., "Balancing Hydropower and Biodiversity."

33. QEI, *Electricity Efficiency through Water Efficiency*, 23–24. (Note: While this is an old study [1992], the energy efficiency of household appliances has greatly increased over the last 2 decades, so the comparison might be even more striking today.) The 19% and 30% figures are from California Energy Commission, "California's Water–Energy Relationship."

34. Mooney and Ehrlich, "Ecosystem Services: A Fragmentary History"; Westman, "How Much Are Nature's Services Worth?"

35. Millennium Ecosystem Assessment, "Living Beyond Our Means: Natural Assets and Human Well-Being," March 2005.

36. Costanza et al., "The Value of the World's Ecosystem Services and Natural Capital." This study's valuation estimates were expressed in 1994 dollars; I used an inflation calculator to express their estimate of $19,580 per hectare in 2016 dollars.

37. Milly et al., "Stationarity Is Dead: Whither Water Management?"

38. Williams et al., "Contribution of Anthropogenic Warming to California Drought during 2012–2014"; Gillis, "California Drought Is Made Worse

by Global Warming, Scientists Say"; most severe drought from Griffin and Anchukaitis, "How Unusual Is the 2012–2014 California Drought?"

39. Schiermeier, "Increased Flood Risk Linked to Global Warming"; "Last Winter's Floods 'Most Extreme on Record in UK,' Says Study," *Guardian*, December 5, 2016.

40. Prein et al., "The Future Intensification of Hourly Precipitation Extremes"; Schlossberg, "Climate Change Will Bring Wetter Storms in U.S., Study Says."

41. Weaver et al., "Preliminary Peak Stage and Streamflow Data at Selected Stream-Gaging Stations in North Carolina and South Carolina for Flooding Following Hurricane Matthew"; Bidgood, "A Wrenching Decision Where Black History and Floods Intertwine."

Chapter 2

1. Alvarez de Williams, "Cocopa"; population of 200 from Luecke et al., *A Delta Once More: Restoring Riparian and Wetland Habitat in the Colorado River Delta*.

2. Juan Butrón and Osvel Hinojosa-Huerta, author interviews, February 4, 2013; expansion and contraction of the Ciénega from Glenn et al., "Ciénega de Santa Clara: Endangered Wetland in the Colorado River Delta, Sonora, Mexico"; current areas of marsh and mudflats from Zamora-Arroyo et al., *Conservation Priorities in the Colorado River Delta, Mexico and the United States*.

3. Osvel Hinojosa-Huerta, Director of Water and Wetlands Program, Pronatura Noroeste, author interview, February 4, 2013.

4. Leopold, *A Sand County Almanac*, 141–49.

5. For more on this Aral Sea experience, see Postel, "Honest Hope."

6. Flow data is for the Southern International Boundary from the International Boundary and Water Commission, El Paso, Texas.

7. Bergman, *Red Delta: Fighting for Life at the End of the Colorado River*, 119.

8. Colorado River Research Group, "The First Step in Repairing the Colorado River's Broken Water Budget: Summary Report"; reservoir levels from the US Bureau of Reclamation at usbr.gov; Scripps Institution of Oceanography, "Lake Mead Could Be Dry by 2021."

9. Luecke et al., *A Delta Once More*, 42.

10. Zamora-Arroyo et al., *Conservation Priorities in the Colorado River Delta, Mexico and the United States.*

11. International Boundary and Water Commission, United States and Mexico, Minute No. 319.

12. Yamilett Carrillo, consultant to Pronatura Noroeste, author interviews during trip to Colorado Delta in February 2013. (Carrillo later became executive director of the Colorado River Delta Water Trust, recently renamed Restauremos El Colorado AC.)

13. Jennifer Pitt, Director, Colorado River Program, Environmental Defense Fund, author interview, Boulder, Colorado, July 16, 2014. (Pitt is now Colorado River Project director for the National Audubon Society.)

14. Howard, "Saving the Colorado River Delta, One Habitat at a Time."

15. Sonoran Institute, "Restoring Mexico's Colorado River Delta, One Tree at a Time."

16. Carrillo, presentation delivered at the 1st Colorado River Basin Water Transactions Workshop.

17. Flessa et al., eds., "Minute 319 Colorado River Limitrophe and Delta Environmental Flows Monitoring Interim Report."

18. Jensen, "Colorado River Delta Flows Help Birds, Plants, Groundwater."

19. Yamilett Carrillo, Executive Director, Restauremos El Colorado AC, San Diego, California, email communication with author, January 12, 2017.

Chapter 3

1. Two percent figure from National Public Radio, "As Brazil's Largest City Struggles with Drought, Residents Are Leaving," November 22, 2015.

2. Postel, "Lessons from São Paulo's Water Shortage."

3. Rocha, "Drought Bites as Amazon's 'Flying Rivers' Dry Up." According to Rocha, meteorologist Jose Marengo first coined the phrase "flying rivers."

4. Salati and Vose, "Amazon Basin: A System in Equilibrium."

5. Castello and Macedo, "Large-Scale Degradation of Amazonian Freshwater Ecosystems."

6. Robbins, "Deforestation and Drought"; Castello and Macedo, previous note, also make the point that by causing shifts in energy and water balances, land-cover changes in the Amazon Basin "may provoke shifts in regional rainfall regimes."

7. This section draws upon author's detailed notes from June 1988 trip to China.

8. Hillel, *Out of the Earth: Civilization and the Life of the Soil*, 173.

9. Zhang Daquan, Chief Engineer, Water and Soil Conservancy, Shaanxi Province, author interview, June 10, 1988.

10. He et al., "Land Use Change and Deforestation on the Loess Plateau."

11. Institute of Management of Loess Plateau, Mizhi Experiment Station, Shaanxi Province, author interviews, June 14, 1988.

12. Data from Shaanxi Control Institute of the Loess Plateau were provided to author during visit to Mizhi County, Shaanxi Province, China, June 1988.

13. Qian et al., "Regional Characteristics of Dust Storms in China"; *World Tibet Network News*, "Death Toll in China Sandstorm Rises to 47," May 11, 1993.

14. Buckingham and Hanson, "Case Example: China Loess Plateau"; Lü et al., "A Policy-Driven Large Scale Ecological Restoration: Quantifying Ecosystem Services Changes in the Loess Plateau of China."

15. Buckingham and Hanson, "Case Example: China Loess Plateau."

16. Ibid.

17. Liu, "Lessons of the Loess Plateau."

18. Buckingham and Hanson, "Case Example: China Loess Plateau."

19. Concerns and payment expiration from Buckingham and Hanson, "Case Example: China Loess Plateau."

20. Hardin, "The Tragedy of the Commons."

21. Shabecoff, "Global Warming Has Begun, Expert Tells Senate."

22. US Forest Service, "The Rising Cost of Wildfire Operations"; Fears, "U.S. Wildfires Just Set an Amazing and Troubling New Record"; Richtel and Santos, "Wildfires, Once Confined to a Season, Burn Earlier and Longer"; Fountain, "Dry Winter and Warm Spring Set Stage for Wildfire in Canada."

23. Basic facts from New Mexico EPSCoR, "New Mexico Fire and Water: Impacts and Lessons Learned from the Las Conchas Fire."

24. Fleck, "Post-Fire Runoff Led to 'Dead Zones.'"

25. Fleck, "Rio Grande Water Restored."

26. Rio Grande Water Fund, "Comprehensive Plan for Wildfire and Water Source Protection."

27. US Forest Service, "The Rising Cost of Wildfire Operations"; Fears, "U.S. Wildfires Just Set an Amazing and Troubling New Record."

28. American Forest Foundation, "Western Water Threatened by Wildfire: It's Not Just a Public Lands Issue."

29. Postel, "Fire and Rain: The One-Two Punch of Flooding after Blazes"; Gartner et al., "Natural Infrastructure: Investing in Forested Landscapes for Source Water Protection in the United States," 101.

30. Postel, "Fire and Rain"; Gartner et al., "Natural Infrastructure," 96, 101.

31. Laura McCarthy, Director of Conservation Programs, The Nature Conservancy–New Mexico, author interview, Albuquerque, New Mexico, February 10, 2016.

32. Lyons, "Rio Grande Water Fund Avoided Cost Analysis."

33. Rio Grande Water Fund, Annual Reports 2015 and 2016.

34. American Water Works Association, "Forest Cover and Its Effects on Water Treatment Costs"; Warziniack et al., "Effect of Forest Cover on Drinking Water Treatment Costs"; Ernst, "Protecting the Source: Land Conservation and the Future of America's Drinking Water."

35. Postel and Thompson, "Watershed Protection: Capturing the Benefits of Nature's Water Supply Services."

36. New York City Department of Environmental Protection, "Land Acquisition," at nyc.gov, viewed on February 23, 2016; Wisnieski, "City's Watershed Protection Plan Seeks Difficult Balance Upstate."

37. Quote in Wisnieski, "City's Watershed Protection Plan Seeks Difficult Balance Upstate."

38. Postel and Thompson, "Watershed Protection: Capturing the Benefits of Nature's Water Supply Services"; Bajak, "In Water We Trust."

39. Whelan, "Liquid Asset."

Chapter 4

1. Recent scientific discoveries suggest this mythological story may align with a historical event; see Normile, "Massive Flood May Have Led to China's Earliest Empire."

2. 1,500 floods from Hillel, *Out of the Earth: Civilization and the Life of the Soil*, 173; Davies, "Central China Floods 1931."

3. Risk Management Solutions, "Central Europe Flooding, August 2002: Event Report."

4. Belgrade from Pearce, *When the Rivers Run Dry: Water—the Defining Cri-*

sis of the Twenty-First Century, 283; European Commission, "Flood Protection: Commission Proposes Concerted EU Action."

5. International Commission for the Protection of the Danube River (ICPDR), "The Danube River Basin: Facts and Figures"; 100-year flood from ICPDR, "Flood Risk Management Plan for the Danube River Basin District," at icpdr.org; *New Haven Register*, "Danube Flooding Affects Ten Nations, Millions of People."

6. Costanza et al., "Changes in the Global Value of Ecosystem Services."

7. Anonymous (by request), "The Downstream Impacts of Ethiopia's Gibe III Dam"; for more on the downstream impacts of dams, see Richter et al., "Lost in Development's Shadow: The Downstream Human Consequences of Dams."

8. Vastag and Sellers, "Floods along the Mississippi River Lead to Renewed Calls for a Change in Strategy"; polder project from Kimmelman, "Going with the Flow."

9. European Parliament, "Environment: European Parliament and Council Reach Agreement on the New Floods Directive."

10. Pearce, "A Successful Push to Restore Europe's Long-Abused Rivers."

11. ICPDR, "Flood Risk Management Plan for the Danube River Basin District."

12. ICPDR, "Ten Years of the Green Corridor."

13. $260 billion from US Executive Office of the President, "Fact Sheet: Taking Action to Protect Communities and Reduce the Cost of Future Flood Disasters"; Robertson and Blinder, "As Louisiana Floodwaters Recede, the Scope of Disaster Comes into View"; US National Oceanic and Atmospheric Administration, National Weather Service, Hydrometeorological Design Studies Center, "Exceedance Probability Analysis for Selected Storm Events," at noaa.gov, viewed August 17, 2016.

14. History of US Army Corps of Engineers at http://www.usace.army.mil/About/History, viewed August 8, 2016.

15. Mississippi River Commission and US Army Corps of Engineers, "Room for the River: Summary Report of the 2011 Mississippi River Flood and Successful Operation of the Mississippi River & Tributaries System."

16. Allen, "Army Corps Makes Tough Calls with Floods."

17. Mississippi River Commission and U.S. Army Corps of Engineers, "Room for the River."

18. Lower Mississippi River Conservation Committee (LMRCC), "Restoring America's Greatest River: A Habitat Restoration Plan for the Lower Mississippi River."

19. Current status from Mississippi River Trust, "Restoring Mississippi River Bottomland Hardwood Forests," at mississippirivertrust.org, viewed August 2, 2016.

20. LMRCC, "Restoring America's Greatest River."

21. The Nature Conservancy, "The Conservancy Reconnected Miles of Floodplain Forest Back to Louisiana's Ouachita River," at nature.org, viewed August 1, 2016.

22. Wetland losses from Wires et al., Upper Mississippi Valley/Great Lakes Waterbird Conservation Plan.

23. Pinter, "One Step Forward, Two Steps Back on U.S. Floodplains."

24. Poff et al., "Sustainable Water Management Under Future Uncertainty with Eco-Engineering Decision Scaling"; Dodge, "Researchers: Building Better Dams Starts with Ecological Insights"; N. LeRoy Poff, Professor of Biology, Colorado State University, author interview, Fort Collins, Colorado, April 18, 2016.

25. Pall et al., "Anthropogenic Greenhouse Gas Contribution to Flood Risk in England and Wales in Autumn 2000."

26. National Academies of Sciences, Engineering and Medicine, *Attribution of Extreme Weather Events in the Context of Climate Change*, 99–100.

27. Ibid., 101–2.

28. Poff et al., "Sustainable Water Management Under Future Uncertainty with Eco-Engineering Decision Scaling."

29. California State Association of Counties (CSAC), "Napa County, 'Innovative Flood Control'—National County Government Month 2015."

30. National Research Council, *Valuing Ecosystem Services: Toward Better Environmental Decision-Making*; $366 million figure from Killam, "Sacramento District Project Wins Public Works Project of the Year"; Techel quote from CSAC, "Napa County, Innovative Flood Control."

31. Robbins, "In Napa Valley, Future Landscapes Are Viewed in the Past."

32. Opperman et al., "Sustainable Floodplains through Large-Scale Reconnection to Rivers"; March 2016 event and six years in ten from Bienick, "Yolo Bypass Floods for 1st Time Since 2012."

33. Yolo Basin Foundation, "About the Yolo Bypass Wildlife Area," at yoloba-

sin.org, viewed August 7, 2016.

Chapter 5

1. *Arab News*, "Almarai Acquires Huge Farmland in Arizona"; National Public Radio, "Saudi Hay Farm in Arizona Tests State's Supply of Groundwater."

2. Postel, *Pillar of Sand*, 78.

3. Giordano, "Global Groundwater? Issues and Solutions"; United Nations Food and Agriculture Organization (FAO), Aquastat Database.

4. Saudi land purchases from Rulli et al., "Global Land and Water Grabbing."

5. Ninety-six percent is from Shiklomanov, *World Water Resources: An Appraisal for the 21st Century.*

6. India from Postel, *Pillar of Sand*, 56–57; 40 percent from FAO, *The State of the World's Land and Water Resources for Food and Agriculture—Managing Systems at Risk*, 38.

7. Morgan, *Water and the Land: A History of American Irrigation.*

8. Stanton et al., *Selected Approaches to Estimate Water-Budget Components of the High Plains, 1940 through 1949 and 2000 through 2009.*

9. US Geological Survey, "Irrigation Causing Declines in the High Plains Aquifer."

10. Postel, *Pillar of Sand*, 80.

11. Wada et al., "Global Depletion of Groundwater Resources"; Konikow, "Contribution of Global Groundwater Depletion since 1900 to Sea-Level Rise."

12. Castle et al., "Groundwater Depletion during Drought Threatens Future Water Security of the Colorado River Basin."

13. University of California–Irvine, "Parched West Is Using Up Underground Water, UCI, NASA Find."

14. Rodell et al., "Satellite-Based Estimates of Groundwater Depletion in India"; Famiglietti et al., "Satellites Measure Recent Rates of Groundwater Depletion in California's Central Valley."

15. Famiglietti, "The Global Groundwater Crisis."

16. Galbraith, "Texas Farmers Battle Ogallala Pumping Limits."

17. Postel, "That Sinking Feeling about Groundwater in Texas"; meters and

enforcement policies from Carmon McCain, Information and Education Supervisor, High Plains Underground Water Conservation District, Lubbock, Texas, email communication with author, April 5, 2016.

18. Rogers, *Historic Reclamation Projects*; National Groundwater Association, "Managed Aquifer Recharge: A Water Supply Management Tool."

19. Rate of subsidence from Boxall, "Overpumping of Central Valley Groundwater Creating a Crisis, Experts Say."

20. Ibid.

21. Garone, *The Fall and Rise of the Wetlands of California's Great Central Valley.*

22. Nick Blom, almond farmer, author interview, Modesto, California, March 23, 2016.

23. Kerlin, "Flooding Farms in the Rain to Restore Groundwater"; 2016 monitoring results from Helen Dahlke, assistant professor of integrated hydrological sciences, University of California–Davis, email communication with author, Davis, California, December 8, 2016.

24. Helen Dahlke, assistant professor of integrated hydrological sciences, University of California–Davis, author interview, Davis, California, March 21, 2016.

25. Two million figure from Public Policy Institute of California, "California's Water"; additional pumping from Howitt et al., "Economic Analysis of the 2015 Drought for California Agriculture."

26. Gabriele Ludwig, director, Sustainability and Environmental Affairs, Almond Board of California, author interview conducted at Nick Blom's farm, Modesto, California, March 23, 2016.

27. Andrew Fahlund, senior program officer, California Water Foundation, Sacramento, California, author interview (by phone), March 18, 2016.

28. Kate Williams, consultant and former program manager with the California Water Foundation, author interview, Sacramento, California, March 22, 2016.

29. RMC Water and Environment, "Creating an Opportunity: Groundwater Recharge through Winter Flooding of Agricultural Land in the San Joaquin Valley."

30. Daniel Mountjoy, director of resource stewardship, Sustainable Conservation, author interview, Modesto, California, March 23, 2016; cost of

dam from Natural Resources Defense Council, "Regional Water Supply Solutions Generally More Cost-Effective than New Dams and Reservoirs."

31. Mountjoy, author interview.

32. Madera Irrigation District, "Madera Irrigation District Continues Water Deliveries and Massive Recharge Effort"; Rodriguez, "Farmers and Water Districts Hope Storm Runoff Can Help Replenish Underground Supplies."

33. Almond Board of California, "Potential for Groundwater Recharge in Almonds"; Disclaimer: WhiteWave is a sponsor of Change the Course, the national restoration initiative I helped launch.

34. Groundwater extraction figures from National Groundwater Association, "Facts about Global Groundwater Usage"; number of wells from Shah, "Groundwater Governance and Irrigated Agriculture."

35. Shah, *Taming the Anarchy: Groundwater Governance in South Asia*, 2.

36. Suutari and Marten, "Water Warriors: Rainwater Harvesting to Replenish Underground Water (Rajasthan, India).

37. Agarwal and Narain, eds., *Dying Wisdom: Rise, Fall and Potential of India's Traditional Water Harvesting Systems*, 344; Suutari and Marten, "Water Warriors."

38. Ten thousand *johads* from Tarun Bharat Sangh at tarunbharatsangh.in, viewed March 9, 2016; one thousand villages from Stockholm International Water Institute (SIWI), "Rajendra Singh—the Water Man of India Wins 2015 Stockholm Water Prize."

39. SIWI, Rajendra Singh interview. (Note: I reordered the sentences in this quote, but this did not alter the meaning.)

40. World Bank Group, "Basin Based Approach for Groundwater Management: Neemrana, District of Alwar, Rajasthan, India."

41. Postel, "Solar Electricity Buybacks May Reduce Groundwater Depletion in India."

42. Ibid.; Gujarat only state from Shah, "India's Irrigation Challenge: Is PMKSY Equal to It?"

Chapter 6

1. For a rich account of the Dust Bowl, see Egan, *The Worst Hard Time.*

2. One liter per calorie from Molden, ed., *Water for Food, Water for Life: A Comprehensive Assessment of Water Management in Agriculture*, 5; eight times from Shiklomanov, "World Fresh Water Resources."

3. Tallman, "No-Till Case Study, Brown's Ranch: Improving Soil Health Improves the Bottom Line"; Ohlson, *The Soil Will Save Us*, 78–112.

4. Brown, "Holistic Regeneration of Our Lands."

5. Rawls et al., "Effect of Soil Organic Carbon on Soil Water Retention"; 58 percent carbon from US Department of Agriculture, Natural Resources Conservation Service, "Soil Quality for Environmental Health," at http://soilquality.org/indicators/total_organic_carbon.html.

6. Blanco-Canqui et al., "Soil Organic Carbon: The Value to Soil Properties"; Jones, "Soil Carbon: Can It Save Agriculture's Bacon?"

7. Loss of organic carbon from Teague et al., "The Role of Ruminants in Reducing Agriculture's Carbon Footprint in North America."

8. Voth, "Tighty Whities Can Tell You about Your Soil Health."

9. Tallman, "No-Till Case Study, Brown's Ranch."

10. Yield and organic matter from Tallman, "No-Till Case Study, Brown's Ranch."

11. Erosion and fertilizer runoff from Ohlson, *The Soil Will Save Us*, 107–8; Strom, "Cover Crops, a Farming Revolution with Deep Roots in the Past."

12. Moebius-Clune, "Introducing the New Soil Health Division," US Department of Agriculture, Natural Resources Conservation Service, webinar, January 12, 2016, available at conservationwebinars.net; Averett, "Healthy Ground, Healthy Atmosphere: Recarbonizing the Earth's Soils."

13. 105 million hectares from Winterbottom et al., "Improving Land and Water Management."

14. Postel, "Getting More Crop Per Drop"; share irrigated from United Nations Food and Agriculture Organization (FAO), Aquastat Database, at fao.org; 200 million from Winterbottom and Reij, "Farmer Innovation: Improving Africa's Food Security through Land and Water Management."

15. Winterbottom et al., "Improving Land and Water Management."

16. McWilliams, "All Sizzle and No Steak: Why Allan Savory's TED Talk about How Cattle Can Reverse Global Warming Is Dead Wrong."

17. Robbins, *Diet for a New America*, 367; Chapagain and Hoekstra, *Water Footprints of Nations: Volume 1: Main Report*, 42. More recently, WFN recalculated the average water footprint of beef to total 15,400 liters per kilogram (1,840 gallons per pound). At that rate a quarter-pounder would consume 460 gallons of water. See Water Footprint Network (WFN), "Product Gallery," at waterfootprint.org/en/resources/interactive-tools/product-gallery/.

18. Beckett and Oltjen, "Estimation of the Water Requirement for Beef Production in the United States"; eggs and coffee from WFN, Product Gallery, viewed October 21, 2016.

19. Ruechel, *Grass-Fed Cattle: How to Produce and Market Natural Beef*, 9.

20. Robert Potts, president, Dixon Water Foundation, author interview, Fort Davis and Marfa, Texas, October 25, 2016.

21. Mankad, "Does Grass-Fed Beef Have Any Heart-Health Benefits That Other Types of Beef Don't?"

22. Species decline from the North American Bird Conservation Initiative, US Committee, *The State of the Birds 2014*; Beth Bardwell, director of conservation, Audubon New Mexico, author interview, Los Lunas, New Mexico, October 17, 2016, and email correspondence, January 5, 2017.

23. Beth Bardwell, author interview and email correspondence.

24. Goodloe, "Ranching and the Practice of Watershed Conservation"; in Loeffler and Loeffler, eds., *Thinking Like a Watershed: Voices from the West*, 193–214.

25. Goodloe quote from Western Landowners Alliance, "Stewardship with Vision: Caring for New Mexico's Streams."

26. Teague et al., "Multi-paddock Grazing on Rangelands: Why the Perceptual Dichotomy between Research Results and Rancher Experience?"; Park et al., "Evaluating the Ranch and Watershed Scale Impacts of Using Traditional and Adaptive Multi-Paddock Grazing on Runoff, Sediment and Nutrient Losses in North Texas, USA."

27. Teague et al., "The Role of Ruminants in Reducing Agriculture's Carbon Footprint in North America."

28. Loss of rangeland from National Science and Technology Council, "The State and Future of U.S. Soils: Framework for a Federal Strategic Plan for Soil Science."

29. Fairlie, *Meat: A Benign Extravagance.*

Chapter 7

1. International Desalination Association, "Desalination by the Numbers," at idadesal.org, viewed November 19, 2016.

2. Rockland Water Coalition, website at sustainablerockland.org; Hellauer, "Sustainability Saturday: How the Rockland Desalination War Was Won."

3. Laurie Seeman, member of Rockland Water Coalition and founding director of Strawtown Studios, author interview (phone), November 21, 2016.

4. Berger, "Desalination Plan Draws Ire in Rockland County"; New York City, Department of Environmental Protection, Water Demand Management Plan.

5. Berger, "Desalination Plan Draws Ire in Rockland County."

6. Amy Vickers, president, Amy Vickers & Associates, Inc., author interview, Amherst, MA, June 23, 2016; 7 billion gallons from Vickers and Bracciano, "Low-Volume Plumbing Fixtures Achieve Water Savings."

7. DeOreo et al., *Residential End Uses of Water, Version 2.*

8. Amy Vickers & Associates, Inc., *Water Losses and Customer Water Use in the United Water New York System.*

9. Ibid., cartoon appears in *Our Town*, "Leaks Are Us: United Water's Fuzzy Data."

10. Rockland County, Office of the Legislature, "Independent Report: Millions of Gallons of Water Available to Rockland County—Rockland Task Force Files Major Water Study with State," press release, New City, New York, July 28, 2015.

11. State of New York, Public Service Commission, Case 16-W-0130, Statement of Harriet Cornell, 13–29.

12. Rockland County, Office of the County Executive, press release, New City, NY, June 27, 2016.

13. Rockland Water Coalition, "As PSC Decision Looms on Rockland

Water Rate Hikes and a Flawed Conservation Plan, Residents Question Whether a Private Corporation Should Manage Public Water."

14. Two percent from American Association for the Advancement of Science, *Atlas of Population & Environment*, at atlas.aaas.org, viewed November 21, 2016; McDonald et al., "Water on an Urban Planet: Urbanization and the Reach of Urban Water Infrastructure."

15. Plumbing code amendment from Vickers, "Water-Use Efficiency Standards for Plumbing Fixtures: Benefits of National Legislation"; data for drop in water use from Stephen Estes-Smargiassi, director of planning, the Massachusetts Water Resources Authority, Boston, Massachusetts, email communication with author, October 18, 2010; average annual demand shows little change since 2010, as evidenced by data at the water authority's website, mwra.state.ma.us, viewed November 21, 2016.

16. Garrick and Hall, "Water Security and Society: Risks, Metrics, and Pathways."

17. PUB, Singapore's National Water Agency, website at www.pub.gov.sg, viewed November 21–22, 2016.

18. Tang, "From Open Sewage to High-Tech Hydrohub, Singapore Leads Water Revolution."

19. Co-siting from Carolyn Khew, "Fifth Singapore Desalination Plant in the Pipeline."

20. Schneider, "In Water-Scarce Regions Desalination Plants Are Risky Investments."

21. Turner et al., "Managing Drought: Learning from Australia."

22. Ibid.

23. "Chicago Green: Roofs" at wendycitychicago.com.

24. US Executive Office of the President, "Commitments to Action on Building a Sustainable Water Future"; Richard G. Luthy, presentation for webinar "Using Graywater and Stormwater to Enhance Local Water Supplies: An Assessment of Risks, Costs and Benefits," hosted by WateReuse, March 10, 2016.

25. WaterWorld, "Kansas City Water's Green Infrastructure Projects Win Sustainability Award"; KC Water Services, "Kansas City's Overflow Control Program: Middle Blue River Basin Green Solutions Pilot Project Final Report."

26. Glen Abrams, director of Sustainable Communities, Pennsylvania Hor-

ticultural Society, presentation for webinar "Training a Green Infrastructure Workforce," hosted by Urban Waters Learning Network, with River Network and Groundwork USA, December 8, 2015. For a comprehensive portfolio of rainwater-harvesting structures and techniques, see Lancaster, *Rainwater Harvesting for Drylands and Beyond*.

27. Cathcart-Keays, "Why Copenhagen Is Building Parks That Can Turn into Ponds."

28. Steinbock, "High Time to Reduce Costs of Floods"; growth in cities from Shepard, "Can 'Sponge Cities' Solve China's Urban Flooding Problem?"

29. Shepard, "Can 'Sponge Cities' Solve China's Urban Flooding Problem?"; funding from Leach, "Soak It Up: China's Ambitious Plan to Solve Urban Flooding with 'Sponge Cities.'"

30. Government target from Leach, "Soak It Up."

31. Katherine Yuhas, manager, Water Resources Division, Albuquerque–Bernalillo County Water Utility Authority, author interviews and field excursion, Albuquerque, New Mexico, November 17, 2016.

32. Kernodle et al., *Simulation of Ground-Water Flow in the Albuquerque Basin, Central New Mexico, 1901–1994, with Projections to 2020*; Albuquerque–Bernalillo County Water Utility Authority, website at abcwua.org, viewed November–December 2016.

33. Southern Nevada Water Authority, website at snwa.com, viewed December 2, 2016.

Chapter 8

1. US Army et al., *Potential Military Chemical/Biological Agents and Compounds*.

2. Pond and watershed areas from Friends of Georgica Pond Foundation, at friendsofgeorgicapond.org, viewed September 6, 2016.

3. Suffolk County Department of Health Services et al., "Investigation of Fish Kills Occurring in the Peconic River–Riverhead, N.Y."

4. New York State Department of Environmental Conservation, "Harmful Algal Blooms and Marine Biotoxins"; Hattenrath et al., "The Influence of Anthropogenic Nitrogen Loading and Meteorological Conditions on the Dynamics and Toxicity of *Alexandrium fundyense* Blooms in a New York (USA) Estuary."

5. Garvey et al., "Opportunity & Critical Path for Water Technology Innovation."

6. Diaz and Rosenberg, "Spreading Dead Zones and Consequences for Marine Ecosystems"; Diaz, "Overview of Hypoxia around the World"; Peconic Bay from Semple, "Long Island Sees a Crisis as It Floats to the Surface."

7. Milman, "Florida Declares State of Local Emergency over Influx of 'God-awful' Toxic Algae"; NASA image from Parker, "Slimy Green Beaches May Be Florida's New Normal"; microcystin levels from Carr, "Tests Reveal Florida's Toxic Algae Is Threatening Not Only the Water Quality but Also the Air."

8. Circle of Blue and the Everglades Foundation, "A State of Emergency: Toxic Algal Blooms Choke Water Supplies in Florida and Around the Globe," remarks during live interactive broadcast, July 21, 2016.

9. Wines, "Behind Toledo's Water Crisis, a Long Troubled Lake Erie."

10. International Fertilizer Industry Association, "Nitrogen Fertilizer Nutrient Consumption," electronic database at www.fertilizer.org.

11. Davis remarks made during Circle of Blue, "A State of Emergency: Toxic Algal Blooms Choke Water Supplies in Florida and around the Globe."

12. Sawyer et al., "Continental Patterns of Submarine Groundwater Discharge Reveal Coastal Vulnerabilities"; Gorder, "Study Maps Hidden Water Pollution in U.S. Coastal Areas."

13. Jennifer Garvey, associate director, New York State Center for Clean Water Technology, Stony Brook University, Stony Brook, New York, author interview conducted in East Hampton, New York, August 27, 2016.

14. Suffolk County (NY), "Comprehensive Water Resources Management 2015: Findings, Recommendations, and Next Steps."

15. Garvey, email communication, September 26, 2016.

16. Suffolk County (NY) Departments of Economic Development & Planning, Health Services, and Public Works, "Advanced Wastewater & Transfer of Development Rights Tour Summary."

17. Commercial and recreational values are from US National Oceanic and Atmospheric Administration (NOAA), "NOAA-Supported Scientists Find Large Gulf Dead Zone, but Smaller than Predicted."

18. White et al., "Nutrient Delivery from the Mississippi River to the Gulf of Mexico and Effects of Cropland Conservation."

19. Mississippi River/Gulf of Mexico Watershed Nutrient Task Force, *2015 Report to Congress.*

20. NOAA, "Average 'Dead Zone' for Gulf of Mexico Predicted."

21. Ibid.; NOAA, "NOAA and Partners Cancel Gulf Dead Zone Summer Cruise."

22. NOAA, "NOAA and Partners Cancel Gulf Dead Zone Summer Cruise."

23. Strom, "Cover Crops, a Farming Revolution with Deep Roots in the Past."

24. Bryant et al., *Counting Cover Crops*; Maryland from Strom, "Cover Crops, a Farming Revolution."

25. McIsaac et al., "Illinois River Nitrate-Nitrogen Concentrations and Loads: Long-Term Variation and Association with Watershed Nitrogen Inputs"; updated finding and quote from University of Illinois, College of Agricultural, Consumer and Environmental Sciences, "Illinois River Water Quality Improvement Linked to More Efficient Corn Production."

26. Murphy et al., *Nitrate in the Mississippi River and Its Tributaries.*

27. Rabotyagov et al., "Cost-Effective Targeting of Conservation Investments to Reduce the Northern Gulf of Mexico Hypoxic Zone."

28. Wetland loss from Wires et al., *Upper Mississippi Valley/Great Lakes Waterbird Conservation Plan.*

29. Groh et al., "Nitrogen Removal and Greenhouse Gas Emissions from Constructed Wetlands Receiving Tile Drainage Water."

30. University of Illinois, College of Agricultural, Consumer and Environmental Sciences, "Wetlands Continue to Reduce Nitrates."

31. Mississippi River/Gulf of Mexico Hypoxia Task Force, "Hypoxia Task Force Success Stories," at epa.gov, viewed September 26, 2016; 6 percent from Miller, "Building Wetlands for Clean Drinking Water"; Mitsch et al., "Reducing Nitrogen Loading to the Gulf of Mexico from the Mississippi River Basin: Strategies to Counter a Persistent Ecological Problem."

32. Environmental Working Group, "Dead in the Water."

Chapter 9

1. Lahnsteiner and Lempert, "Water Management in Windhoek, Namibia."

2. For more on Orange County's operation, see Schwartz, "Water Flowing from Toilet to Tap May Be Hard to Swallow."

3. Stuart White, director, Institute for Sustainable Futures, University Technology Sydney, presentation for the Alliance for Water Efficiency, "Managing Drought: Learning from Australia," webinar, May 2, 2016.

4. Kurt Dahl, managing director, Permeate Partners, Sydney, New South Wales, Australia, communication during author's visit to Pennant Hills Golf Club, June 2010.

5. GE Water and Process Technologies, "Australian Golf Course Recycles Municipal Wastewater with Onsite ZeeWeed MBR."

6. Dahl and Kirby, "Sewer Mining as an Alternative Water Source—the Pennant Hills Experience."

7. Allen, presentation at "EcoDistricts Summit: Membrane Technology for District-Scale Wastewater Treatment."

8. Jacobs, "Keep Your Mouth Closed: Aquatic Olympians Face a Toxic Stew"; discharge to oceans from National Research Council, "Water Reuse: Potential for Expanding the Nation's Water Supply through Reuse of Municipal Wastewater," 1.

9. Houston, Texas, example from National Research Council, "Water Reuse: Potential for Expanding the Nation's Water Supply through Reuse of Municipal Wastewater," 26.

10. Cooley et al., "Water Reuse Potential in California: Issue Brief."

11. Rogoway, "Apple's Newest Innovation: Wastewater Treatment to Cool Prineville Data Centers."

12. Siegel, *Let There Be Water: Israel's Solution for a Water-Starved World*, 81–89.

13. Shuval et al., "Wastewater Irrigation in Developing Countries: Health Effects and Technical Solutions."

14. Melbourne Water, website at melbournewater.com.au.

15. Thebo et al., "Global Assessment of Urban and Peri-urban Agriculture: Irrigated and Rainfed Croplands."

16. Monterey Regional Water Pollution Control Agency, "Turning Wastewater into Safe Water."

17. Monterey Regional Water Pollution Control Agency, "Monterey Wastewater Reclamation for Agriculture"; quote from video at WateReuse .org/Pentair-and-water-sector-groups-collaborate-to-advance-sustainable-water-reuse-in agriculture/, May 24, 2016.

18. Assouline and Narkis, "Effects of Long-Term Irrigation with Treated Wastewater on the Hydraulic Properties of a Clayey Soil."

19. Lautze et al., *Global Experiences in Water Reuse.*

20. International Water Management Institute, "First Global Estimate of Urban Agriculture Reveals Area Size of the EU That's Boosting Food Security in Cities"; 10 percent of global food production from Lautze et al., *Global Experiences in Water Reuse.*

21. Keraita et al., *On-farm Treatment Options for Wastewater, Greywater and Fecal Sludge, with Special Reference to West Africa.*

22. This section draws upon the author's visit to Las Arenitas, Baja California, Mexico, in February 2013, as well as interviews with engineers on-site, Javier Orduño Valdez, director of the state public services commission in Mexicali, and with Osvel Hinojosa-Huerta, director of water and wetlands, Pronatura Noroeste, Ensenada, Baja California, Mexico.

23. Bird count from Sonoran Institute, "From Sewage to Sanctuary: Las Arenitas Treatment Wetland," at sonoraninstitute.org, viewed May 6, 2016.

Chapter 10

1. Singler, "If You Remove It They Will Come—Restoring Amethyst Brook, MA."

2. Amy Singler and Brian Graber, associate director and senior director, respectively, of river restoration, American Rivers, author interviews, Northampton, Massachusetts, January 2014.

3. Poff et al., "The Natural Flow Regime: A Paradigm for River Conservation and Restoration."

4. US Geological Survey, "Concepts for National Assessment of Water Availability and Use," says there are nearly 77,000 dams, which I adjusted downward to account for recent dam removals.

5. Waldman, *Running Silver: Restoring Atlantic Rivers and Their Great Fish Migrations*, 4–5.

6. Nedeau, *Freshwater Mussels and the Connecticut River Watershed*, 5.

7. American Rivers, "72 Dams Removed in 2016, Improving Safety for River Communities"; Elwha removals from Mapes, "Elwha: Roaring Back to Life."

8. Opperman, "Fish Run Through It: The Importance of Maintaining and Reconnecting Free-Flowing Rivers"; The University of Maine et al., "After More than a Century, Endangered Shortnose Sturgeon Find Historic Habitat Post Dam Removal."

9. Schneider, "Popularity of Big Hydropower Projects Diminishes around the World."

10. Ibid.

11. Present number from International Commission on Large Dams, World Register of Dams, at http://www.icold-cigb.org/GB/World_register/general_synthesis.asp, viewed January 21, 2017.

12. For more on the history and development of environmental flows, see Arthington, *Environmental Flows: Saving Rivers in the Third Millennium*; also Postel and Richter, *Rivers for Life: Managing Water for People and Nature.*

13. Warner et al., "Restoring Environmental Flows through Adaptive Reservoir Management: Planning, Science, and Implementation through the Sustainable Rivers Project."

14. The Nature Conservancy, "Modernizing Dam Operations," and "Modernizing Water Management: Building a National Sustainable Rivers Program."

15. Texas Parks and Wildlife Department and Texas Fish and Wildlife Conservation Office, "Final Report for Big Cypress Bayou Paddlefish Reintroduction Assessment; Caddo Lake Institute, "The Paddlefish Experiment and Education Project."

16. Warner et al., "Restoring Environmental Flows"; quote from TNC, "An Ancient Fish Returns to Caddo Lake," video at www.nature.org/paddle fish, viewed May 19, 2016.

17. Skinner and Langford, "Legislating for Sustainable Basin Management: The Story of Australia's Water Act (2007)"; tripling of extractions and low Murray flows from Blackmore, "The Murray–Darling Basin Cap on Diversions—Policy and Practice for the New Millennium."

18. Australian Bureau of Meteorology, "Recent Rainfall, Drought and Southern Australia's Long-term Rainfall Decline," at bom.gov.au, viewed April 12, 2016; Draper, "Australia's Dry Run."

19. Murray cod from Wahlquist, *Thirsty Country: Options for Australia*, 157; losses to Ngarrindjeri from Draper, "Australia's Dry Run."

20. Blackmore, "The Murray–Darling Basin Cap."

21. Skinner and Langford, "Legislating for Sustainable Basin Management."

22. Murray–Darling Basin Authority, "Guide to the Proposed Basin Plan"; Cooper, "Murray–Darling Plan Gets Fiery Reception."

23. Quote from Cullen, "Burke Unveils Final Murray–Darling Plan."

24. Young et al., "Science Review of the Estimation of an Environmentally Sustainable Level of Take for the Murray–Darling Basin"; Skinner and Langford, "Legislating for Sustainable Basin Management."

25. Power, "Peak Water: Aquifers and Rivers Are Running Dry. How Three Regions Are Coping"; Coleambally Irrigation website at new.colyirr.com .au, and video at youtube.com/watch?v=Rb4Dw271000, viewed May 2016.

26. Australian Bureau of Meteorology, "Record Rainfall and Widespread Flooding," at www.bom.gov.au/climate/enso/history/ln-2010-12/rainfall -flooding.shtml, viewed on November 9, 2016.

27. Vidot and Worthington, "Murray–Darling Basin Authority Recommends Reducing Water Buybacks in Northern Communities."

28. This section draws upon Postel, "How Smarter Irrigation Might Save Rare Mussels and Ease a Water War," and upon interviews conducted August 11, 2016, with Calvin Perry, Stripling Irrigation Research Park, University of Georgia, Camilla; George Vellidis, Crop and Soil Sciences Department, University of Georgia, Tifton; and Casey Cox, executive director, Flint River Partnership.

29. Rugel et al., "Effects of Irrigation Withdrawals on Streamflows in a Karst Environment: Lower Flint River Basin, Georgia, USA"; Golladay et al., "Stream Habitat and Mussel Populations Adjacent to AAWCM Sites in the Lower Flint River Basin: Project Report."

30. Golladay et al., "Stream Habitat and Mussel Populations."

31. Stephen Golladay, aquatic biologist, J. W. Jones Ecological Research Center, Newton, Georgia, author interview (phone), August 15, 2016.

32. Postel and Richter, *Rivers for Life*, 190.

33. Ibid.; Supreme Court of the United States, *State of Florida v. State of Georgia*, Report of the Special Master, quote p. 31.

34. Vellidis et al., "A Dynamic Variable Rate Irrigation Control System."

Chapter 11

1. Garcia, *Early Phoenix*, 16; Tempe Historical Society, website at tempehis toricalsociety.org, viewed April 2, 2012.

2. Doyel, "Hohokam Cultural Evolution in the Phoenix Basin"; Fish and Fish, "Hohokam Political and Social Organization."

3. Woodhouse et al., "A 1,200-Year Perspective of 21st Century Drought in Southwestern North America."

4. Fish and Fish, eds., *The Hohokam Millennium.*

5. Vano et al., "Understanding Uncertainties in Future Colorado River Streamflow."

6. Beard, *Deadbeat Dams; Why We Should Abolish the US Bureau of Reclamation and Tear Down Glen Canyon Dam*; Lustgarten, "Unplugging the Colorado River."

7. Fisher, "Stream Ecosystems of the Western United States."

8. Center for Biological Diversity, "San Pedro River," Tucson, Arizona, at biologicaldiversity.org, viewed January 14, 2017.

9. Arizona Land and Water Trust, "Water Restoration Monitoring Report Form," Tucson, Arizona, October 18, 2016. Provided through email communication with author.

10. Gori et al., "Gila River Flow Needs Assessment."

11. US Bureau of Reclamation, "Appraisal Level Report on the Arizona Water Settlements Act Tier-2 Proposals and Other Diversion and Storage Configurations."

12. Western Resource Advocates, "Filling the Gap: Meeting Future Urban and Domestic Water Needs in Southwestern New Mexico."

13. Gila Conservation Coalition, "ISC and NM CAP Entity Release Gila River Diversion Options; Gaume, Testimony to the New Mexico Interstate Stream Commission.

14. Gila Conservation Coalition, website at gilaconservation.org.

15. Daniel P. Beard, former commissioner, US Bureau of Reclamation, public presentation, Albuquerque, NM, November 18, 2015.

16. Martha S. Cooper, Southwest New Mexico Program, The Nature Conservancy, email communication with author, April 4, 2016.

17. Much of this section derives from interviews during the author's visit to the Verde Valley from July 28 to August 1, 2013, and from Postel, "Arizona Irrigators Share Water with Desert River."

18. Number of mammal and bird species from Haney et al., "Ecological Implications of Verde River Flows."

19. Haney et al., 54.

20. Report from Kim Schonek, Verde River Project Manager, The Nature Conservancy–Arizona, to the Bonneville Environmental Foundation, November 12, 2015; 2016 update from Todd Reeve, CEO, Bonneville Environmental Foundation, Portland, OR, email correspondence with author, January 18, 2017.

21. Postel, "Two Arizona Vineyards Give Back to a River through a Voluntary Water Exchange."

22. Loomis, "$20 Million Plan to Aid Arizona's Stressed-Out Verde River."

Chapter 12

1. Simulation and broadcast at Deborah Byrd, "This Date in Science: Earthrise."

2. US National Aeronautics and Space Administration (NASA), "Apollo 8: Christmas at the Moon"; 1 billion from Greenspan, "Remembering the Apollo 8 Christmas Eve Broadcast."

3. McKibben, *Eaarth: Making a Life on a Tough New Planet*.

4. This section draws upon Postel, "How the Yampa River, and Its Dependents, Survived the Drought of 2012."

5. Shuttleworth, "Agreement Entitles Whanganui River to Legal Identity"; Postel, "A River in New Zealand Gets a Legal Voice."

6. Kendall, "A New Law of Nature"; Vidal, "Bolivia Enshrines Natural World's Rights with Equal Status for Mother Earth."

7. Stone, *Should Trees Have Standing? Toward Legal Rights for Natural Objects*; US Supreme Court, *Sierra Club v. Morton* (1972); Postel, "The Missing Piece: A Water Ethic."

8. This section draws upon Postel, "A Dam, Dying Fish, and a Montana Farmer's Lifelong Quest to Right a Wrong."

9. Opening of habitat from French, "Bypassing the Barrier."

10. Leopold, *A Sand County Almanac*, 203; Postel, *Last Oasis*, 183–91.

11. Bratter, "Sustainability Straight Talk: Working across the Public and Private Sector to Solve the Global Water Crisis."

Bibliography

Books

Agarwal, Anil, and Sunita Narain, eds. *Dying Wisdom: Rise, Fall and Potential of India's Traditional Water Harvesting Systems*. New Delhi: Centre for Science and Environment, 1997.

Arthington, Angela H. *Environmental Flows: Saving Rivers in the Third Millennium*. Berkeley and Los Angeles, CA: University of California Press, 2012.

Barnett, Cynthia. *Rain: A Natural and Cultural History*. New York: Crown Publishers, 2015.

Beard, Daniel P. *Deadbeat Dams: Why We Should Abolish the U.S. Bureau of Reclamation and Tear Down Glen Canyon Dam*. Boulder, CO: Johnson Books, 2015.

Bergman, Charles. *Red Delta: Fighting for Life at the End of the Colorado River*. Golden, CO: Fulcrum Publishing, 2002.

Brown, Peter G., and Jeremy J. Schmidt, eds. *Water Ethics: Foundational Readings for Students and Professionals*. Washington, DC: Island Press, 2010.

Childs, Craig. *The Secret Knowledge of Water*. New York: Back Bay Books, 2001.

Cushing, C.E., K.W. Cummins, and G.W. Minshall, eds. *River and Stream Ecosystems of the World*. Berkeley, CA: University of California Press, 1995, 2006.

Daily, Gretchen C., ed. *Nature's Services: Societal Dependence on Natural Ecosystems*. Washington, DC: Island Press, 1997.

deBuys, William. *A Great Aridness: Climate Change and the Future of the American Southwest*. New York: Oxford University Press, 2011.

Egan, Timothy. *The Worst Hard Time: The Untold Story of Those Who Survived the Great American Dust Bowl*. Boston, MA: Houghton Mifflin Harcourt, 2006.

Fairlie, Simon. *Meat: A Benign Extravagance*. White River Junction, VT: Chelsea Green Publishing, 2010.

Fausch, Kurt D. *For the Love of Rivers: A Scientist's Journey*. Corvallis, OR: Oregon State University Press, 2015.

Fish, Suzanne K., and Paul R. Fish, eds. *The Hohokam Millennium*. Santa Fe, NM: School for Advanced Research Press, 2007.

Fishman, Charles. *The Big Thirst: The Secret Life and Turbulent Future of Water*. New York: Free Press, 2011.

Fleck, John. *Water Is for Fighting Over—and Other Myths about Water in the West*. Washington, DC: Island Press, 2016.

France, Robert Lawrence, ed. *Thoreau on Water: Reflecting Heaven*. Boston, New York: Houghton Mifflin Company, 2001.

Garcia, Kathleen. *Early Phoenix*. Mt. Pleasant, SC: Arcadia Publishing, 2008.

Garone, Philip. *The Fall and Rise of the Wetlands of California's Great Central Valley*. Oakland, CA: University of California Press, 2011.

Glennon, Robert. *Unquenchable: America's Water Crisis and What to Do About It*. Washington, DC: Island Press, 2009.

Gumerman, George J., ed. *Exploring the Hohokam: Prehistoric Desert Peoples of the American Southwest*. Albuquerque, NM: University of New Mexico Press, 1991.

Hillel, Daniel J. *Out of the Earth: Civilization and the Life of the Soil*. New York: The Free Press, 1991.

Kandel, Robert. *Water from Heaven: The Story of Water from the Big Bang to the Rise of Civilization and Beyond*. New York: Columbia University Press, 2003.

Kingsolver, Barbara. *Small Wonder*. New York: HarperCollins Publishers, 2002.

Kolbert, Elizabeth. *The Sixth Extinction: An Unnatural History*. New York: Henry Holt and Company, 2014.

Lancaster, Brad. *Rainwater Harvesting for Drylands and Beyond*. Tucson, AZ: Rainsource Press, 2013.

Leopold, Aldo. *A Sand County Almanac*. New York: Oxford University Press, 1949.

Leopold, Luna B. *A View of the River*. Cambridge, MA: Harvard University Press, 1994.

Loeffler, Jack, and Celestia Loeffler, eds. *Thinking Like a Watershed: Voices from the West*. Albuquerque: University of New Mexico Press, 2012.

McKibben, Bill. *Eaarth: Making a Life on a Tough New Planet*. New York: St. Martin's Griffin, 2010, 2011.

McNamee, Gregory. *Gila: The Life and Death of an American River*. Albuquerque: University of New Mexico Press, 1998.

McNeill, J.R. *Something New Under the Sun: An Environmental History of the Twentieth-Century World*. New York: W.W. Norton & Co, 2000.

McPhee, John. *The Control of Nature*. New York: Farrar Straus Giroux, 1989.

Molden, David, ed. *Water for Food, Water for Life: A Comprehensive Assessment of Water Management in Agriculture*. London: Earthscan, with International Water Management Institute, Colombo, 2007.

Morgan, Robert M. *Water and the Land: A History of American Irrigation*. Fairfax, VA: The Irrigation Association, 1993.

Nabhan, Gary Paul. *Growing Food in a Hotter, Drier Land*. White River Junction, VT: Chelsea Green Publishing, 2013.

Nedeau, Ethan Jay. *Freshwater Mussels and the Connecticut River Watershed*. Greenfield, MA: Connecticut River Watershed Council, 2008.

Nichols, Wallace J. *Blue Mind: The Surprising Science That Shows How Being Near, In, On, or Under Water Can Make You Happier, Healthier, More Connected, and Better at What You Do*. New York: Little, Brown and Company, 2014.

Niman, Nicolette Hahn. *Defending Beef: The Case for Sustainable Meat Production*. White River Junction, VT: Chelsea Green Publishing, 2014.

Ohlson, Kristin. *The Soil Will Save Us: How Scientists, Farmers and Foodies Are Healing the Soil to Save the Planet*. New York: Rodale, 2014.

Oliver, Mary. *Long Life: Essays and Other Writings*. Cambridge, MA: Da Capo Press, 2004.

Pearce, Fred. *When the Rivers Run Dry: Water—the Defining Crisis of the Twenty-First Century*. Boston, MA: Beacon Press, 2006.

Pollan, Michael. *In Defense of Food: An Eater's Manifesto*. New York, NY: Penguin Books, 2008.

Postel, Sandra. *Last Oasis: Facing Water Scarcity*. New York: W.W. Norton & Co., 1992, 1997.

Postel, Sandra. *Pillar of Sand: Can the Irrigation Miracle Last?* New York: W.W. Norton & Co, 1999.

Postel, Sandra, and Brian Richter. *Rivers for Life: Managing Water for People and Nature.* Washington, DC: Island Press, 2003.

Radkau, Joachim. *Nature and Power: A Global History of the Environment.* New York: Cambridge University Press, 2008.

Richter, Brian. *Chasing Water: A Guide for Moving from Scarcity to Sustainability.* Washington, DC: Island Press, 2014.

Robbins, John. *Diet for a New America.* Tiburon, CA: H. J. Kramer, 1987.

Ruechel, Julius. *Grass-Fed Cattle: How to Produce and Market Natural Beef.* North Adams, MA: Storey Publishing, 2006.

Savoy, Lauret Edith. *Trace: Memory, History, Race, and the American Landscape.* Berkeley, CA: Counterpoint Press, 2015.

Schwartz, Judith D. *Cows Save the Planet.* White River Junction, VT: Chelsea Green Publishing, 2013.

Schwartz, Judith D. *Water in Plain Sight: Hope for a Thirsty World.* New York: St. Martin's Press, 2016.

Shah, Tushaar. *Taming the Anarchy: Groundwater Governance in South Asia.* Washington, DC: Resources for the Future, 2009.

Siegel, Seth M. *Let There Be Water: Israel's Solution for a Water-Starved World.* New York: Thomas Dunne Books, 2015.

Solomon, Steven. *Water: The Epic Struggle for Wealth, Power, and Civilization.* New York: Harper, 2010.

Stone, Christopher. *Should Trees Have Standing? Toward Legal Rights for Natural Objects.* New York: Avon Books, 1974. Original in 1972.

US National Research Council. *Valuing Ecosystem Services: Toward Better Environmental Decision-Making.* Washington, DC: The National Academy Press, 2005.

Vickers, Amy. *Handbook of Water Use and Conservation: Homes, Landscapes, Businesses, Industries, Farms.* Amherst, MA: WaterPlow Press, 2001.

Wahlquist, Åsa. *Thirsty Country: Options for Australia.* Crows Nest, New South Wales: Jacana Books, 2008.

Waldman, John. *Running Silver: Restoring Atlantic Rivers and Their Great Fish Migrations.* Guilford, CT: Lyons Press, 2013.

Wulf, Andrea. *The Invention of Nature: Alexander von Humboldt's New World.* New York: Alfred A. Knopf, 2015.

Journal Articles, Book Chapters, Conference Papers, Reports

Allen, Chris. "Presentation at EcoDistricts Summit: Membrane Technology for District-Scale Wastewater Treatment." Portland, Oregon, October 26–28, 2011.

Alvarez de Williams, Anita. "Cocopa." In Alfonso Ortiz, ed., *Handbook of North American Indians*, Vol. 10 (Southwest). Washington, DC: Smithsonian, 1983.

American Forest Foundation. "Western Water Threatened by Wildfire: It's Not Just a Public Lands Issue." Washington, DC: American Forest Foundation, 2015.

American Water Works Association. "Forest Cover and Its Effects on Water Treatment Costs." *AWWA Connections*, August 3, 2016. https://www.awwa .org/publications/connections/connections-story/articleid/4275/forest -cover-and-its-effect-on-water-treatment-costs.aspx.

Amy Vickers & Associates, Inc. *Water Losses and Customer Water Use in the United Water New York System*. Report prepared for Rockland County Task Force on Water Resources Management. Rockland County, NY: July 2015.

Anonymous (by request). "The Downstream Impacts of Ethiopia's Gibe III Dam." Report. Berkeley, CA: International Rivers, January 2013.

Assouline, S., and K. Narkis. "Effects of Long-Term Irrigation with Treated Wastewater on the Hydraulic Properties of a Clayey Soil." *Water Resources Research* 47, no. 8 (2011). https://doi.org/10.1029/2011WR010498.

Averett, Nancy. "Healthy Ground, Healthy Atmosphere: Recarbonizing the Earth's Soils." *Environmental Health Perspectives* 124, no. 2 (2016): A30–A35. https://doi.org/10.1289/ehp.124-A30.

Bajak, Aleszu. "In Water We Trust." *Ensia*, July 14, 2014.

Beckett, J.L., and J.W. Oltjen. "Estimation of the Water Requirement for Beef Production in the United States." *Journal of Animal Science* 71 (1993): 818–26.

Blackmore, Don J. "The Murray–Darling Basin Cap on Diversions—Policy and Practice for the New Millennium." *National Water* (June 1999): 1–12.

Blanco-Canqui, Humberto, Charles A. Shapiro, Charles S. Wortmann, et al. "Soil Organic Carbon: The Value to Soil Properties." *Journal of Soil and Water Conservation* 68, no. 5 (September–October 2013): 129A–34A. https://doi.org/10.2489/jswc.68.5.129A.

Bryant, Lara, Ryan Stockwell, and Trisha White. *Counting Cover Crops*. Merrifield, VA: National Wildlife Federation, 2013.

Buckingham, Kathleen, and Craig Hanson. "Case Example: China Loess Plateau." In *The Restoration Diagnostic: A Method for Developing Forest Landscape Restoration Strategies by Rapidly Assessing the Status of Key Success Factors*, by Craig Hanson, Kathleen Buckingham, Sean DeWitt, and Lars Laestadius. Report. Washington, DC: World Resources Institute, 2015. http://www.wri.org/publication/restoration-diagnostic.

California Energy Commission. "California's Water–Energy Relationship." Staff Report. Sacramento, CA: November 2005.

Carrillo, Yamilett. Presentation delivered at the 1st Colorado River Basin Water Transactions Workshop, Grand Junction, Colorado, September 7, 2016.

Castello, Leandro, and Marcia N. Macedo. "Large-Scale Degradation of Amazonian Freshwater Ecosystems." *Global Change Biology* 22, no. 3 (2016): 990–1007. https://doi.org/10.1111/gcb.13173.

Castle, Stephanie L., Brian F. Thomas, John T. Reager, et al. "Groundwater Depletion during Drought Threatens Future Water Security of the Colorado River Basin." *Geophysical Research Letters* 41, no. 16 (2014): 5904–11. https://doi.org/10.1002/2014GL061055.

Chapagain, A.K., and A.Y. Hoekstra. *Main Report, Value of Water Research Report Series 16*. Vol. 1. Water Footprints of Nations. Delft, the Netherlands: UNESCO, 2004.

Colorado River Research Group. "The First Step in Repairing the Colorado River's Broken Water Budget: Summary Report." December 2014.

Cooley, Heather, Peter Gleick, and Robert Wilkinson, "Water Reuse Potential in California: Issue Brief." Oakland, CA: Pacific Institute, and San Francisco, CA: Natural Resources Defense Council, June 2014.

Costanza, Robert, Ralph d'Arge, Rudolf de Groot, et al. "The Value of the World's Ecosystem Services and Natural Capital." *Nature* 387, no. 6630 (1997): 253–60. https://doi.org/10.1038/387253a0.

Costanza, Robert, Rudolf de Groot, Paul Sutton, et al. "Changes in the Global Value of Ecosystem Services." *Global Environmental Change* 26 (2014): 152–8. https://doi.org/10.1016/j.gloenvcha.2014.04.002.

Dahl, Kurt, and Richard Kirby. "Sewer Mining as an Alternative Water Source —the Pennant Hills Experience." *Australian Turfgrass Management,* 50–54.

https://www.agcsa.com.au/sites/default/files/uploaded-content/website
-content/atm-journal/Water%20Management%20-%20Sewer%20Min
ing,%20the%20Pennant%20Hills%20Experience.pdf.

DeOreo, William B., Peter Mayer, Benedykt Dziegielewski, et al. *Residential End Uses of Water, Version 2*. Denver, CO: Water Research Foundation, 2016.

Diaz, Robert J. "Overview of Hypoxia around the World." *Journal of Environmental Quality* 30, no. 2 (2001): 275–81. https://doi.org/10.2134/jeq2001 .302275x.

Diaz, Robert J., and Rutger Rosenberg. "Spreading Dead Zones and Consequences for Marine Ecosystems." *Science* 321, no. 5891 (2008): 926–9. https://doi.org/10.1126/science.1156401.

Doyel, David E. "Hohokam Cultural Evolution in the Phoenix Basin." In *Exploring the Hohokam: Prehistoric Desert Peoples of the American Southwest*, ed. George J. Gumerman, 231–78. Albuquerque, NM: University of New Mexico Press, 1991.

Draper, Robert. "Australia's Dry Run." *National Geographic* (April 2009). http:// ngm.nationalgeographic.com/2009/04/murray-darling/draper-text.

D'Souza, Rohan. "Framing India's Hydraulic Crisis: The Politics of the Modern Large Dam." *Monthly Review* (New York, NY) 60, no. 3 (July–August 2008): 112. https://doi.org/10.14452/MR-060-03-2008-07_7.

Ernst, Caryn. *Protecting the Source: Land Conservation and the Future of America's Drinking Water*. San Francisco, CA: Trust for Public Land, 2004.

Famiglietti, J.S. "The Global Groundwater Crisis." *Nature Climate Change* 4, no. 11 (2014): 945–8. https://doi.org/10.1038/nclimate2425.

Famiglietti, J.S., M. Lo, S.L. Ho, et al. "Satellites Measure Recent Rates of Groundwater Depletion in California's Central Valley." *Geophysical Research Letters* 38, no. 3 (2011). https://doi.org/10.1029/2010GL046442.

Feaster, Toby D., John M. Shelton, and Jeanne C. Robbins, "Preliminary Peak Stage and Streamflow Data at Selected USGS Stream-Gaging Stations for the South Carolina Flood of October 2015." US Geological Survey Open-File Report 2015–1201. Reston, VA: November 2015.

Fish, Paul R., and Suzanne K. Fish. "Hohokam Political and Social Organization." In *Exploring the Hohokam: Prehistoric Desert Peoples of the American Southwest*, ed. George J. Gumerman, 151–75. Albuquerque, NM: University of New Mexico Press, 1991.

Fisher, Stuart G. "Stream Ecosystems of the Western United States." In *River and Stream Ecosystems of the World*, ed. C.E. Cushing, K.W. Cummins, and G.W. Minshall, 61–87. Berkeley, CA: University of California Press, 2006.

Flessa, Karl, Eloise Kendy, and Karen Schlatter, eds. "Minute 319 Colorado River Limitrophe and Delta Environmental Flows Monitoring Interim Report." Prepared for the International Boundary and Water Commission, May 19, 2016.

French, Brett. "Bypassing the Barrier." *Montana Outdoors*, May–June 2008. http://fwp.mt.gov/mtoutdoors/HTML/articles/2008/BypassingtheBarrier.htm.

Garrick, Dustin, and Jim W. Hall. "Water Security and Society: Risks, Metrics, and Pathways." *Annual Review of Environment and Resources* 39, no. 1 (2014): 611–39. https://doi.org/10.1146/annurev-environ-013012-09 3817.

Gartner, Todd, James Mulligan, Rowan Schmidt, and John Gunn. "Natural Infrastructure: Investing in Forested Landscapes for Source Water Protection in the United States." Washington, DC: World Resources Institute, n.d.

Garvey, Jennifer, Harold W. Walker, and Christopher J. Gobler. "Opportunity & Critical Path for Water Technology Innovation." Presentation at the Strategic Planning Symposium, Center for Clean Water Technology, Stony Brook University, June 23, 2016.

Gaume, Norman. Testimony to the New Mexico Interstate Stream Commission, Tucumcari, New Mexico, April 30, 2014.

Giordano, Mark. "Global Groundwater? Issues and Solutions." *Annual Review of Environment and Resources* 34, no. 1 (2009): 153–78. https://doi.org/10 .1146/annurev.environ.030308.100251.

Glenn, Edward P., Richard S. Felger, Alberto Búrquez, et al. "Ciénega de Santa Clara: Endangered Wetland in the Colorado River Delta, Sonora, Mexico." *Natural Resources Journal* 32 (1992): 817–24.

Golladay, Stephen, David W. Hicks, Nathalie Smith, and Brian Clayton. "Stream Habitat and Mussel Populations Adjacent to AAWCM Sites in the Lower Flint River Basin: Project Report." Newton, GA: J. W. Jones Ecological Research Center, October 10, 2014.

Goodloe, Sid. "Ranching and the Practice of Watershed Conservation." In *Thinking Like a Watershed: Voices from the West*, ed. Jack Loeffler and Celestia Loeffler, 193–214. Albuquerque, NM: University of New Mexico Press, 2012.

Gori, D., M.S. Cooper, E.S. Soles, et al., "Gila River Flow Needs Assessment." Report. Arlington, VA: The Nature Conservancy, July 2014.

Griffin, Daniel, and Kevin J. Anchukaitis. "How Unusual Is the 2012–2014 California Drought?" *Geophysical Research Letters* 41, no. 24 (2014): 9017–23. https://doi.org/10.1002/2014GL062433.

Groh, T.A., L.E. Gentry, and M.B. David. "Nitrogen Removal and Greenhouse Gas Emissions from Constructed Wetlands Receiving Tile Drainage Water." *Journal of Environmental Quality* 44, no. 3 (2015): 1001–10. https://doi.org/10.2134/jeq2014.10.0415.

Haney, J.A., D.S. Turner, A.E. Springer, et al. "Ecological Implications of Verde River Flows." Report. Arizona Water Institute, The Nature Conservancy, and the Verde River Basin Partnership, 2008. azconservation .org/dl/TNCAZ_VerdeRiver_Ecological_Flows.pdf.

Hardin, Garrett. "The Tragedy of the Commons." *Science* 162, no. 3859 (1968): 1243–8. https://doi.org/10.1126/science.162.3859.1243.

Hattenrath, Theresa K., Donald M. Anderson, and Christopher J. Gobler. "The Influence of Anthropogenic Nitrogen Loading and Meteorological Conditions on the Dynamics and Toxicity of *Alexandrium fundyense* Blooms in a New York (USA) Estuary." *Harmful Algae* 9, no. 4 (2010): 402–12. https://doi.org/10.1016/j.hal.2010.02.003.

He, Jing-Feng, Jin-Hong Guan, and Wen-Hui Zhang. "Land Use Change and Deforestation on the Loess Plateau." In *Restoration and Development of the Degraded Loess Plateau, China*, ed. Atsushi Tsunekawa, Guobin Liu, Norikazu Yamanaka, et al., 111–20. Tokyo, Japan: Springer, 2014. https:// doi.org/10.1007/978-4-431-54481-4_8.

Hoekstra, Arjen Y., and Mesfin M. Mekonnen. "The Water Footprint of Humanity." *Proceedings of the National Academy of Sciences of the United States of America* 109, no. 9 (2012): 3232–7. https://doi.org/10.1073/pnas .1109936109.

Howitt, Richard E., Duncan MacEwan, Josué Medellín-Azuara, et al. *Economic Analysis of the 2015 Drought for California Agriculture*. Davis, CA: Center for Watershed Sciences, University of California–Davis, 2015.

International Boundary and Water Commission, United States and Mexico. "Minute No. 319: Interim International Cooperative Measures in the Colorado River Basin through 2017 and Extension of Minute 318 Cooperative Measures to Address the Continued Effects of the April 2010 Earthquake in the Mexicali Valley, Baja California." Coronado, California, November 20, 2012.

International Commission for the Protection of the Danube River. Flood Risk Management Plan for the Danube River Basin District. Vienna, Austria, 2015. https://www.icpdr.org/main/sites/default/files/nodes/documents/1stdfrmp-final.pdf.

International Commission for the Protection of the Danube River. "Ten Years of the Green Corridor." *Danube Watch*, January 2010.

Jackson, Robert B., Stephen R. Carpenter, Clifford N. Dahm, et al. "Water in a Changing World." *Issues in Ecology*, no. 9. Washington, DC: Ecological Society of America, Spring 2001. https://doi.org/10.1890/1051 -0761(2001)011[1027:WIACW]2.0.CO;2.

Jacobsen, Thorkild, and Robert M. Adams. "Salt and Silt in Ancient Mesopotamian Agriculture." *Science* 128, no. 3334 (1958): 1251–58.

Jaramillo, Fernando, and Georgia Destouni. "Local Flow Regulation and Irrigation Raise Global Human Water Consumption and Footprint." *Science* 350, no. 6265 (2015): 1248–51. https://doi.org/10.1126/science.aad1010.

Jones, Christine. "Soil Carbon: Can It Save Agriculture's Bacon?" Agriculture & Greenhouse Emissions Conference, Australian Farm Institute, Adelaide, South Australia, May 18–19, 2010.

KC Water Services. "Kansas City's Overflow Control Program: Middle Blue River Basin Green Solutions Pilot Project Final Report." Kansas City, MO: November 2013.

Keraita, B., P. Drechsel, A. Klutse, et al. *On-Farm Treatment Options for Wastewater, Greywater and Fecal Sludge, with Special Reference to West Africa.* Colombo, Sri Lanka: International Water Management Institute, 2014. https://doi.org/10.5337/2014.203.

Kernodle, John Michael, Douglas P. McAda, and Condé R. Thorn. *Simulation of Ground-Water Flow in the Albuquerque Basin, Central New Mexico, 1901–1994, with Projections to 2020.* Water-Resources Investigations Report 94–4251. Albuquerque, NM: US Geological Survey, Water Resources Division, 1995.

Konikow, Leonard F. "Contribution of Global Groundwater Depletion since 1900 to Sea-Level Rise." *Geophysical Research Letters* 38, no. 17 (2011). https://doi.org/10.1029/2011GL048604.

Lahnsteiner, J., and G. Lempert. "Water Management in Windhoek, Namibia." *Water Science and Technology* 55, nos. 1–2 (2007): 441–8. https://doi.org 10.2166/wst.2007.022.

Lautze, Jonathan, Emilie Stander, Pay Drechsel, et al. *Global Experiences in Water Reuse. Report of the CGIAR Research Program on Water, Land and Ecosystems.* Colombo, Sri Lanka: International Water Management Institute, 2014.

Lesk, Corey, Pedram Rowhani, and Navin Ramankutty. "Influence of Extreme Weather Disasters on Global Crop Production." *Nature* 529, no. 7584 (January 7, 2016): 84–87. https://doi.org/10.1038/nature16467.

Lower Mississippi River Conservation Committee. "Restoring America's Greatest River: A Habitat Restoration Plan for the Lower Mississippi River." Vicksburg, MS: Lower Mississippi River Conservation Committee, 2015. http://www.lmrcc.org/wp-content/uploads/2015/04/RARG-FINAL-4-21-2015.pdf.

Lü, Yihe, Bojie Fu, Xiaoming Feng, et al. "A Policy-Driven Large Scale Ecological Restoration: Quantifying Ecosystem Services Changes in the Loess Plateau of China." *PLoS One* 7, no. 2 (February 16, 2012): e31782. https://doi.org/10.1371/journal.pone.0031782.

Luecke, Daniel F., Jennifer Pitt, Chelsea Congdon, et al. *A Delta Once More: Restoring Riparian and Wetland Habitat in the Colorado River Delta.* Washington, DC: Environmental Defense Fund, 1999.

Lyons, Dale. "Rio Grande Water Fund Avoided Cost Analysis." Arlington, VA: The Nature Conservancy, July 2015.

McDonald, Robert I., Katherine Weber, Julie Padowski, et al. "Water on an Urban Planet: Urbanization and the Reach of Urban Water Infrastructure." *Global Environmental Change* 27 (2014): 96–105. https://doi.org/10.1016/j.gloenvcha.2014.04.022.

McGee, William J. "Water as a Resource." In *Water Ethics: Foundational Readings for Students and Professionals*, ed. Peter G. Brown and Jeremy J. Schmidt, 87–90. Washington, DC: Island Press, 2010. (Note: The full, original work of the same title by McGee was published in *American Academy of Political and Social Science* 33, no. 3 [1909]: 37–50.)

McIsaac, Gregory F., Mark B. David, and George Z. Gertner. "Illinois River Nitrate-Nitrogen Concentrations and Loads: Long-Term Variation and Association with Watershed Nitrogen Inputs." *Journal of Environmental*

Quality 45, no. 4 (2016): 1268–75. https://doi.org/10.2134/jeq2015.10 .0531.

Millennium Ecosystem Assessment. "Living Beyond Our Means: Natural Assets and Human Well-Being." Statement from the Board, March 2005. http://www.wri.org/publication/millennium-ecosystem-assessment-liv ing-beyond-our-means.

Milly, P.C.D., Julio Betancourt, Malin Falkenmark, et al. "Stationarity Is Dead: Whither Water Management?" *Science* 319, no. 5863 (2008): 573–4. https://doi.org/10.1126/science.1151915.

Mississippi River Commission and US Army Corps of Engineers. "Room for the River: Summary Report of the 2011 Mississippi River Flood and Successful Operation of the Mississippi River & Tributaries System." Vicksburg, MS: December 2012.

Mississippi River/Gulf of Mexico Watershed Nutrient Task Force. *2015 Report to Congress*. Washington, DC: US Environmental Protection Agency, August 2015.

Mitsch, William J., John W. Day, Jr., J. Wendell Gilliam, et al. "Reducing Nitrogen Loading to the Gulf of Mexico from the Mississippi River Basin: Strategies to Counter a Persistent Ecological Problem." *BioScience* 51, no. 5 (2001): 373–88. https://academic.oup.com/bioscience/article/51/5/373 /243987/Reducing-Nitrogen-Loading-to-the-Gulf-of-Mexico?related -urls=yes&legid=bioscience;51/5/373&cited-by=yes&legid=biosci ence;51/5/373.

Monterey Regional Water Pollution Control Agency, "Monterey Wastewater Reclamation for Agriculture." Report prepared by Engineering Science, Berkeley, California, April 1987.

Mooney, Harold A., and Paul R. Ehrlich. "Ecosystem Services: A Fragmentary History." In *Nature's Services: Societal Dependence on Natural Ecosystems*, ed. Gretchen C. Daily, 11–19. Washington, DC: Island Press, 1997.

Murphy, Jennifer C., Robert M. Hirsch, and Lori A Sprague. *Nitrate in the Mississippi River and Its Tributaries, 1980–2010: An Update*. Reston, VA: US Geological Survey, 2013. https://doi.org/10.3133/sir20135169.

Murray–Darling Basin Authority. *Guide to the Proposed Basin Plan*. Canberra, Australian Capital Territory, 2010.

National Academies of Sciences, Engineering, and Medicine. *Attribution of*

Extreme Weather Events in the Context of Climate Change. Washington, DC: The National Academies Press, 2016.

National Groundwater Association. "Facts about Global Groundwater Usage." Westerville, OH, August 2015.

National Groundwater Association. "Managed Aquifer Recharge: A Water Supply Management Tool." Westerville, OH. http://www.ngwa.org/Media -Center/briefs/Documents/Managed-Aquifer-Recharge.pdf.

National Research Council. *Water Reuse: Potential for Expanding the Nation's Water Supply through Reuse of Municipal Wastewater.* Washington, DC: The National Academies Press, 2012.

National Science and Technology Council. "The State and Future of U.S. Soils: Framework for a Federal Strategic Plan for Soil Science." Washington, DC: Office of the President of the United States, December 2016.

Natural Resources Defense Council. "Regional Water Supply Solutions Generally More Cost-Effective than New Dams and Reservoirs." Fact Sheet. July 2014.

New Mexico EPSCoR. *New Mexico Fire and Water: Impacts and Lessons Learned from the Las Conchas Fire.* Albuquerque, NM: University of New Mexico, 2012.

New York City, Department of Environmental Protection. Water Demand Management Plan. Circa 2015.

Normile, Dennis. "Massive Flood May Have Led to China's Earliest Empire." *Science*, August 4, 2016. https://doi.org/10.1126/science.aag0729.

North American Bird Conservation Initiative. *US Committee. The State of the Birds 2014.* Washington, DC: US Department of the Interior, 2014.

Opperman, Jeffrey J., Gerald E. Galloway, Joseph Fargione, et al. "Sustainable Floodplains through Large-Scale Reconnection to Rivers." *Science* 326, no. 5959 (2009): 1487–8. https://doi.org/10.1126/science.1178256.

Pall, Pardeep, Tolu Aina, Daithi A. Stone, et al. "Anthropogenic Greenhouse Gas Contribution to Flood Risk in England and Wales in Autumn 2000." *Nature* 470, no. 7334 (2011): 382–5. https://doi.org/10.1038/nature 09762.

Park, Jong-Yoon, Srinivasulu Ale, W. Richard Teague, et al. "Evaluating the Ranch and Watershed Scale Impacts of Using Traditional and Adaptive

Multi-Paddock Grazing on Runoff, Sediment and Nutrient Losses in North Texas, USA." *Agriculture, Ecosystems & Environment* 240 (2017): 32–44. https://doi.org/10.1016/j.agee.2017.02.004.

Pearce, Fred. "A Successful Push to Restore Europe's Long-Abused Rivers." *Yale Environment 360*, December 10, 2013. http://e360.yale.edu/features/a_successful_push_to_restore_europes_long-abused_rivers.

Pinter, Nicholas. "One Step Forward, Two Steps Back on U.S. Floodplains." *Science* 308, no. 5719 (2005): 207–8. https://doi.org/10.1126/science.1108411.

Pisani, Donald J. "Water Planning in the Progressive Era: The Inland Waterways Commission Reconsidered." *Journal of Policy History* 18, no. 4 (2006): 389–418. https://doi.org/10.1353/jph.2006.0014.

Poff, N. LeRoy, J. David Allan, Mark B. Bain, et al. "The Natural Flow Regime: A Paradigm for River Conservation and Restoration." *BioScience* 47, no. 11 (1997): 769–84. https://doi.org/10.2307/1313099.

Poff, N. LeRoy, Casey M. Brown, Theodore E. Grantham, et al. "Sustainable Water Management under Future Uncertainty with Eco-Engineering Decision Scaling." *Nature Climate Change* 6 (2016): 25–34. https://doi.org/10.1038/NClimate2765.

Pomeranz, Kenneth. "The Great Himalayan Watershed: Agrarian Crisis, Mega-Dams and the Environment." *New Left Review* 58 (July–August 2009). https://newleftreview.org/II/58/kenneth-pomeranz-the-great-himalayan-watershed.

Population Reference Bureau. "2015 World Population Data Sheet." Washington, DC: Population Reference Bureau, 2015.

Postel, Sandra L. "The Forgotten Infrastructure: Safeguarding Freshwater Ecosystems." *Journal of International Affairs* 61 (2008): 75–90.

Postel, Sandra L. "Getting More Crop Per Drop." In *State of the World: Innovations That Nourish the Planet*, ed. Linda Starke, 39–48. New York: W.W. Norton & Company, 2011.

Postel, Sandra. "Honest Hope." In *Written in Water: Messages of Hope for Earth's Most Precious Resource*, ed. Irena Salina, 46–59. Washington, DC: National Geographic Books, 2010.

Postel, Sandra. "The Missing Piece: A Water Ethic." *American Prospect* 19, no. 6 (May 27, 2008): A14–A15.

Postel, Sandra. "Where Have All the Rivers Gone?" *World Watch* 8, no. 3 (May–June 1995): 9–19.

Postel, Sandra L., Gretchen C. Daily, and Paul R. Ehrlich. "Human Appropriation of Renewable Fresh Water." *Science* 271, no. 5250 (1996): 785–88.

Postel, Sandra L., and Barton H. Thompson, Jr. "Watershed Protection: Capturing the Benefits of Nature's Water Supply Services." *Natural Resources Forum* 29, no. 2 (2005): 98–108. https://doi.org/10.1111/j.1477-8947.2005.00119.x.

Postel, Sandra L., and Aaron T. Wolf. "Dehydrating Conflict." *Foreign Policy*, no. 126 (September–October 2001): 59–67. https://doi.org/10.2307/3183260.

Power, Matthew. "Peak Water: Aquifers and Rivers Are Running Dry. How Three Regions Are Coping." *Wired Magazine*, April 21, 2008. https://www.wired.com/2008/04/ff-peakwater/.

Prein, Andreas F., Roy M. Rasmussen, Kyoko Ikeda, et al. "The Future Intensification of Hourly Precipitation Extremes." *Nature Climate Change* 7, no. 1 (2017): 48–52. https://doi.org/10.1038/nclimate3168.

Propst, David L., Keith B. Gido, and Jerome A. Stefferud. "Natural Flow Regimes, Nonnative Fishes, and Native Fish Persistence in Arid-Land River Systems." *Ecological Applications* 18, no. 5 (2008): 1236–52.

Public Policy Institute of California. "California's Water." Sacramento and San Francisco, CA: April 2015.

QEI, Inc. *Electricity Efficiency through Water Efficiency*. Report for the Southern California Edison Company. Springfield, NJ: QEI, Inc., 1992.

Qian, Weihong, Xu Tang, and Linsheng Quan. "Regional Characteristics of Dust Storms in China." *Atmospheric Environment* 38, no. 29 (2004): 4895–907. https://doi.org/10.1016/j.atmosenv.2004.05.038.

Rabotyagov, Sergey S., Todd D. Campbell, Michael White, et al. "Cost-Effective Targeting of Conservation Investments to Reduce the Northern Gulf of Mexico Hypoxic Zone." *Proceedings of the National Academy of Sciences of the United States of America* 111, no. 52 (2014): 18530–5. https://doi.org/10.1073/pnas.1405837111.

Rawls, W.J., Y.A. Pachepsky, J.C. Ritchie, et al. "Effect of Soil Organic Carbon on Soil Water Retention." *Geoderma* 116, nos. 1–2 (2003): 61–76. https://doi.org/10.1016/S0016-7061(03)00094-6.

Ricciardi, Anthony, and Joseph B. Rasmussen. "Extinction Rates of North

American Freshwater Fauna." *Conservation Biology* 13, no. 5 (1999): 1220–2. https://doi.org/10.1046/j.1523-1739.1999.98380.x.

Richter, B.D., S.L. Postel, C. Revenga, et al. "Lost in Development's Shadow: The Downstream Human Consequences of Dams." *Water Alternatives* 3, no. 2 (2010): 14–42.

Rio Grande Water Fund. "Comprehensive Plan for Wildfire and Water Source Protection." Albuquerque, NM: Rio Grande Water Fund, July 2014. http://nmconservation.org/rgwf/rgwf_compplan.pdf.

Rio Grande Water Fund. "Wildfire and Water Source Protection: Annual Report 2015." https://www.nature.org/ourinitiatives/regions/northamerica/united states/newmexico/new-mexico-rio-grande-water-fund-2015-an nual-report.pdf.

Rio Grande Water Fund. "Wildfire and Water Source Protection: Annual Report 2016." https://www.nature.org/ourinitiatives/habitats/riverslakes /rgwf-2016-annual-report.pdf.

Risk Management Solutions. "Central Europe Flooding, August 2002: Event Report." Newark, CA: Risk Management Solutions, 2003.

RMC Water and Environment. "Creating an Opportunity: Groundwater Recharge through Winter Flooding of Agricultural Land in the San Joaquin Valley." Sacramento and San Francisco, CA: October 2015.

Rodell, Matthew, Isabella Velicogna, and James S. Famiglietti. "Satellite-Based Estimates of Groundwater Depletion in India." *Nature* 460, no. 7258 (2009): 999–1002. https://doi.org/10.1038/nature08238.

Rogers, Jedediah S. *Historic Reclamation Projects*. Washington, DC: US Bureau of Reclamation, 2009.

Rugel, Kathleen, C. Rhett Jackson, J. Joshua Romeis, et al. "Effects of Irrigation Withdrawals on Streamflows in a Karst Environment: Lower Flint River Basin, Georgia, USA." *Hydrological Processes* (2011). https://doi.org /10.1002/hyp.8149.

Rulli, Cristina, Antonio Savori, and Paolo D'Odorico. "Global Land and Water Grabbing." *Proceedings of the National Academy of Sciences of the United States of America* 110, no. 3 (2013): 892–7. https://doi.org/10.1073/pnas .1213163110.

Salati, Eneas, and Peter B. Vose. "Amazon Basin: A System in Equilibrium." *Science* 225, no. 4658 (1984): 129–38. https://doi.org/10.1126/science.225 .4658.129.

Sawyer, Audrey H., Cedric H. David, and James S. Famiglietti. "Continental Patterns of Submarine Groundwater Discharge Reveal Coastal Vulnerabilities." *Science* 353, no. 6300 (2016): 705–7. https://doi.org/10.1126/science.aag1058.

Schiermeier, Quirin. "Increased Flood Risk Linked to Global Warming." *Nature* 470, no. 316 (2011): 16. https://doi.org/10.1038/470316a.

Shah, Tushaar. "Groundwater Governance and Irrigated Agriculture." Report. Stockholm, Sweden: Global Water Partnership, 2014.

Shah, Tushaar. "India's Irrigation Challenge: Is PMKSY Equal to It?" Presentation at the Centre for Policy Research, New Delhi, March 10, 2016.

Shiklomanov, Igor A. "World Fresh Water Resources." In *Water in Crisis: A Guide to the World's Fresh Water Resources*, ed. Peter H. Gleick, 13–24. New York and London: Oxford University Press, 1993.

Shiklomanov, Igor A. *World Water Resources: An Appraisal for the 21st Century*. Paris, France: UNESCO, International Hydrological Programme, 1998.

Shuval, Hillel, Avner Adin, Badri Fattal, et al. *Wastewater Irrigation in Developing Countries: Health Effects and Technical Solutions*. Washington, DC: World Bank, 1986.

Skinner, Dominic, and John Langford. "Legislating for Sustainable Basin Management: The Story of Australia's Water Act (2007)." *Water Policy* 15, no. 6 (2013): 871–94. https://doi.org/10.2166/wp.2013.017.

Stanton, Jennifer S., Sharon L. Qi, Derek W. Ryter, et al. *Selected Approaches to Estimate Water-Budget Components of the High Plains, 1940 through 1949 and 2000 through 2009*. Reston, VA: US Geological Survey, 2011.

State of New York, Public Service Commission. Case 16-W-0130—Proceeding on a motion of the Commission as to the Rates, Charges and Regulations of Suez Water New York Inc. for Water Service. Public Statement Hearing (Transcription of Proceedings). Rockland Community College, Suffern, NY: June 16, 2016.

Suffolk County. "Reclaim Our Water Initiative, Comprehensive Water Resources Management 2015: Findings, Recommendations, and Next Steps." Riverhead, NY: March 23, 2015.

Suffolk County Department of Health Services, New York State Department of Environmental Conservation, and Stony Brook University School of Marine and Atmospheric Sciences. "Investigation of Fish Kills Occurring in the Peconic River–Riverhead, N.Y. Spring 2015." Riverhead, NY: January 2016.

Suffolk County Departments of Economic Development & Planning, Health Services, and Public Works. "Advanced Wastewater & Transfer of Development Rights Tour Summary." Riverhead, NY: April 28, 2014.

Supreme Court of the United States. *Sierra Club v. Morton.* 405 U.S. 727 (1972). https://supreme.justia.com/cases/federal/us/405/727/case.html.

Supreme Court of the United States. *State of Florida v. State of Georgia.* Report of the Special Master (Ralph I. Lancaster, Pierce Atwood LLP, Portland, Maine), February 14, 2017. https://www.pierceatwood.com/sites/default/files/Docket%20636%20Special%20Master%20Report,%20FL%20v%20GA%20No.%20142%20Orig.%20(W6008636x7AC2E).PDF.

Suutari, Amanda, and Gerry Marten. "Water Warriors: Rainwater Harvesting to Replenish Underground Water (Rajasthan, India)." The EcoTipping Points Project: Models for Success in a Time of Crisis, June 2005. http://www.ecotippingpoints.org/our-stories/indepth/india-rajasthan-rainwater-harvest-restoration-groundwater-johad.html.

Syvitski, James P.M., Charles J. Vörösmarty, Albert J. Kettner, et al. "Impact of Humans on the Flux of Terrestrial Sediment to the Global Coastal Ocean." *Science* 308, no. 5720 (2005): 376–80. https://doi.org/10.1126/science.1109454.

Tallman, Susan. "No-Till Case Study, Brown's Ranch: Improving Soil Health Improves the Bottom Line." Butte, MT: National Sustainable Agriculture Information Service, National Center for Appropriate Technology, October 2012.

Teague, Richard, Fred Provenza, Urs Kreuter, et al. "Multi-Paddock Grazing on Rangelands: Why the Perceptual Dichotomy between Research Results and Rancher Experience?" *Journal of Environmental Management* 128 (2013): 699–717. https://doi.org/10.1016/j.jenvman.2013.05.064.

Teague, W.R., S. Apfelbaum, R. Lal, et al. "The Role of Ruminants in Reducing Agriculture's Carbon Footprint in North America." *Journal of Soil and Water Conservation* 71, no. 2 (2016): 156–64. https://doi.org/10.2489/jswc.71.2.156.

Texas Parks and Wildlife Department and Texas Fish and Wildlife Conservation Office. "Final Report for Big Cypress Bayou Paddlefish Reintroduction Assessment February 2014 to May 2015." https://www.fws.gov/southwest/fisheries/txfwco/documents/Big_Cypress_Bayou_Paddlefish_Reintroduction_Assessment_2014-2015.pdf.

Thebo, A.L., P. Drechsel, and E.F. Lambin. "Global Assessment of Urban and Peri-urban Agriculture: Irrigated and Rainfed Croplands." *Environmental Research Letters* 9, no. 11 (2014): 114002. https://doi.org/10.1088/1748-9326/9/11/114002.

Turner, A., S. White, J. Chong, et al. "Managing Drought: Learning from Australia." New South Wales, Australia: University of Technology Sydney, February 2016.

United Nations Food and Agriculture Organization (FAO). Aquastat Database. http://www.fao.org/nr/water/aquastat/countries_regions/saudi_arabia/index.stm.

United Nations Food and Agriculture Organization (FAO). *The State of the World's Land and Water Resources for Food and Agriculture—Managing Systems at Risk*. Rome: FAO and London: Earthscan, 2011.

United Nations Office for Disaster Risk Reduction and Centre for Research on the Epidemiology of Disasters. "The Human Cost of Weather Related Disasters 1995–2015." https://www.unisdr.org/we/inform/publications/46796.

US Army, Marine Corps, Navy, and Air Force. "Potential Military Chemical/Biological Agents and Compounds." Report. Washington, DC: January 2005.

US Bureau of Reclamation. "Appraisal Level Report on the Arizona Water Settlements Act Tier-2 Proposals and Other Diversion and Storage Configurations: Technical Support Provided to the New Mexico Interstate Stream Commission." Phoenix, AZ: US Bureau of Reclamation, Phoenix Area Office, July 2014.

US Executive Office of the President. "Commitments to Action on Building a Sustainable Water Future." Washington, DC: March 22, 2016.

US Forest Service. "The Rising Cost of Wildfire Operations." Washington, DC: US Department of Agriculture, August 2015.

US Geological Survey. "Concepts for National Assessment of Water Availability and Use." Circular 1223. Reston, VA: 2002.

US National Oceanic and Atmospheric Administration, National Centers for Environmental Information. "National Overview—Annual 2016." Silver Spring, MD: US Department of Commerce, 2016.

US National Oceanic and Atmospheric Administration, National Weather Service. "The Record Front Range and Eastern Colorado Floods of September 11–17, 2013." Silver Spring, MD: US Department of Commerce, June 2014.

Vano, Julie A., Bradley Udall, Daniel R. Cayan, et al. "Understanding Uncertainties in Future Colorado River Streamflow." *Bulletin of the American Meteorological Society* 95, no. 1 (January 2014): 59–78. https://doi.org/10.1175/BAMS-D-12-00228.1.

Vellidis, George, Vasileios Liakos, Wesley Porter, et al. "A Dynamic Variable Rate Irrigation Control System." 13th International Conference on Precision Agriculture, St. Louis, Missouri, July 31–August 4, 2016.

Vickers, Amy. "Water-Use Efficiency Standards for Plumbing Fixtures: Benefits of National Legislation." *Journal AWWA* 82, no. 5 (1990): 51–54.

Vickers, Amy, and David Bracciano. "Low-Volume Plumbing Fixtures Achieve Water Savings." *Opflow*, July 2014. https://doi.org/10.5991/OPF.2014.40.0047.

Vörösmarty, Charles, and Dork Sahagian. "Anthropogenic Disturbance of the Terrestrial Water Cycle." *BioScience* 50, no. 9 (2000): 753–65. https://doi.org/10.1641/0006-3568(2000)050[0753:ADOTTW]2.0.CO;2.

Voth, Kathy. "Tighty Whities Can Tell You about Your Soil Health." *On Pasture*, November 21, 2016.

Wada, Yoshihide, Ludovicus P.H. van Beek, Cheryl M. van Kempen, et al. "Global Depletion of Groundwater Resources." *Geophysical Research Letters* 37, no. 20 (2010). https://doi.org/10.1029/2010GL044571.

Warner, Andrew T., Leslie B. Bach, and John T. Hickey. "Restoring Environmental Flows through Adaptive Reservoir Management: Planning, Science, and Implementation through the Sustainable Rivers Project." *Hydrological Sciences Journal* 59, nos. 3–4 (2014): 770–85. https://doi.org/10.1080/02626667.2013.843777.

Warziniack, Travis, Chi Ho Sham, Robert Morgan, et al. "Effect of Forest Cover on Drinking Water Treatment Costs." Report. Denver, CO: American Water Works Association, 2016.

Weaver, J.C., T.D. Feaster, and J.C. Robbins. "Preliminary Peak Stage and

Streamflow Data at Selected Stream-Gaging Stations in North Carolina and South Carolina for Flooding Following Hurricane Matthew." US Geological Survey, Open-File Report 2016–1205. Reston, VA: October 2016.

Western Landowners Alliance. "Stewardship with Vision: Caring for New Mexico's Streams." Santa Fe, NM: Western Landowners Alliance, 2014.

Western Resource Advocates. "Filling the Gap: Meeting Future Urban and Domestic Water Needs in Southwestern New Mexico." Boulder, CO: August 2014.

Westman, Walter E. "How Much Are Nature's Services Worth?" *Science* 197, no. 4307 (1977): 960–4. https://doi.org/10.1126/science.197.4307.960.

Whelan, Carolyn. "Liquid Asset." *Nature Conservancy*, Autumn 2010: 43–49.

White, M.J., C. Santhi, N. Kannan, et al. "Nutrient Delivery from the Mississippi River to the Gulf of Mexico and Effects of Cropland Conservation." *Journal of Soil and Water Conservation* 69, no. 1 (2014): 26–40. https://doi.org/10.2489/jswc.69.1.26.

Williams, A. Park, Richard Seager, John T. Abatzoglou, et al. "Contribution of Anthropogenic Warming to California Drought during 2012–2014." *Geophysical Research Letters* 42, no. 16 (2015): 6819–28. https://doi.org/10.1002/2015GL064924.

Winemiller, K.O., P.B. McIntyre, L. Castello, et al. "Balancing Hydropower and Biodiversity in the Amazon, Congo, and Mekong." *Science* 351, no. 6269 (2016): 128–9. https://doi.org/10.1126/science.aac7082.

Winterbottom, Robert, and Chris Reij. "Farmer Innovation: Improving Africa's Food Security through Land and Water Management." Washington, DC: World Resources Institute, October 16, 2013.

Winterbottom, Robert, Chris Reij, Dennis Garrity, et al. "Improving Land and Water Management." Working Paper. Washington, DC: World Resources Institute, October 2013.

Wires, Linda R., Stephen J. Lewis, Gregory J. Soulliere, et al. "Upper Mississippi Valley/Great Lakes Waterbird Conservation Plan." Report submitted to US Fish and Wildlife Service, Fort Snelling, Minnesota, March 2010.

Woodhouse, Connie A., David M. Meko, Glen M. Macdonald, et al. "A 1,200-Year Perspective of 21st Century Drought in Southwestern North America." *Proceedings of the National Academy of Sciences of the United*

States of America 107, no. 50 (2010): 21283–8. https://doi.org/10.1073/pnas.0911197107.

World Bank Group. "NEW-Basin Based Approach for Groundwater Management: Neemrana, District of Alwar, Rajasthan, India." Washington, DC: World Bank Group, n.d. https://www.waterscarcitysolutions.org/new-basin-based-approach-for-groundwater-management/.

World Economic Forum. "The Global Risks Report 2016." Geneva, Switzerland, 2016. http://www3.weforum.org/docs/GRR/WEF_GRR16.pdf.

Young, W.J., N. Bond, J. Brookes, et al. "Science Review of the Estimation of an Environmentally Sustainable Level of Take for the Murray–Darling Basin." Report to the Murray–Darling Basin Authority from the CSIRO Water for a Healthy Country Flagship, Canberra, Australian Capital Territory, 2011.

Zamora-Arroyo, Francisco, Jennifer Pitt, Steve Cornelius, et al. *Conservation Priorities in the Colorado River Delta, Mexico and the United States.* Prepared by the Sonoran Institute, Environmental Defense Fund, University of Arizona, Pronatura Noroeste, Centro de Investigación en Alimentación y Desarrollo and World Wildlife Fund, 2005.

News Articles, Blogs, Websites

Albuquerque Bernalillo County Water Utility Authority, Albuquerque, NM. http://www.abcwua.org.

Allen, Greg. "Army Corps Makes Tough Calls with Floods." Transcript. National Public Radio, May 9, 2011. http://www.npr.org/2011/05/09/136056393/army-corps-makes-tough-calls-with-floods.

Almond Board of California. "Potential for Groundwater Recharge in Almonds." October 26, 2015. http://www.almonds.com/newsletters/outlook/potential-groundwater-recharge-almonds.

American Association for the Advancement of Science. *Atlas of Population & Environment.* http://atlas.aaas.org/.

American Rivers. "72 Dams Removed in 2016, Improving Safety for River Communities." Washington, DC, February 9, 2016. https://www.americanrivers.org/conservation-resource/72-dams-removed-2016/.

Arab News. "Almarai Acquires Huge Farmland in Arizona." March 11, 2014. http://www.arabnews.com/news/537336.

Associated Press. "Danube Flooding Affects Ten Nations, Millions of People." *New Haven Register*, June 6, 2013. http://www.nhregister.com/article/NH/20130606/NEWS/306069778.

Australian Bureau of Meteorology. "Recent Rainfall, Drought and Southern Australia's Long-Term Rainfall Decline." April 2015. http://www.bom.gov.au/climate/updates/articles/a010-southern-rainfall-decline.shtml.

Australian Bureau of Meteorology. "Record Rainfall and Widespread Flooding." http://www.bom.gov.au/climate/enso/history/ln-2010-12/rainfall-flooding.shtml.

Berger, Joseph. "Desalination Plan Draws Ire in Rockland County." *New York Times*, November 12, 2014. https://www.nytimes.com/2014/11/13/nyregion/desalination-plan-draws-ire-in-rockland-county.html?_r=0.

Bidgood, Jess. "A Wrenching Decision Where Black History and Floods Intertwine." *New York Times*, December 9, 2016. https://www.nytimes.com/2016/12/09/us/princeville-north-carolina-hurricane-matthew-floods-black-history.html.

Bienick, David. "Yolo Bypass to Flood for the First Time." KCRA News, March 13, 2016. http://www.kcra.com/article/yolo-bypass-to-flood-for-the-first-time/6256521.

Boxall, Bettina. "Overpumping of Central Valley Groundwater Creating a Crisis, Experts Say." *Los Angeles Times*, March 18, 2015. http://www.latimes.com/local/california/la-me-groundwater-20150318-story.html.

Bratter, Deanna. "Sustainability Straight Talk: Working Across the Public and Private Sector to Solve the Global Water Crisis." *Huffington Post*, December 12, 2016. http://www.huffingtonpost.com/entry/584eb9b5e4b01713310512cc?timestamp=1481558573566.

British Broadcasting Corporation. "Flood Bill for Winter Estimated at 1.3bn, Says ABI." January 11, 2016. http://www.bbc.com/news/business-35277668.

Brown, Gabe. "Holistic Regeneration of Our Lands." Presentation at the Quivira Coalition Conference, Albuquerque, New Mexico, November 14, 2012. Video of presentation. https://www.youtube.com/watch?v=rKjX3UdVDFU.

Byrd, Deborah. "This Date in Science: Earthrise." http://earthsky.org/space/apollo-8-earthrise-december-24-1968-new-simulation, posted December 24, 2015.

Caddo Lake Institute. "The Paddlefish Experiment and Education Project." http://www.caddolakeinstitute.us/paddlefish_project.html.

California State Association of Counties (NCAC). "Napa County, 'Innovative Flood Control'—National County Government Month 2015." Video. https://www.youtube.com/watch?v=lL44wt4kkpw, posted April 22, 2015.

Carr, Ada. "Tests Reveal Florida's Toxic Algae Is Threatening Not Only the Water Quality but Also the Air." The Weather Channel, July 28, 2016. https://weather.com/science/environment/news/florida-martin-county-algae-toxic-air-particles-marina-rio.

Cathcart-Keays, Athlyn. "Why Copenhagen Is Building Parks That Can Turn into Ponds." Citiscope, January 21, 2016. http://citiscope.org/story/2016/why-copenhagen-building-parks-can-turn-ponds.

Center for Biological Diversity. "San Pedro River." Tucson, Arizona. http://www.biologicaldiversity.org/programs/public_lands/rivers/san_pedro_river/.

City of Fort Collins, Colorado. "2013 Flood." http://www.fcgov.com/utilities/what-we-do/stormwater/flooding/2013-flood/.

Cooper, Hayden. "Murray–Darling Plan Gets Fiery Reception." Transcript. Australian Broadcasting Corporation, October 14, 2010. http://www.abc.net.au/lateline/content/2010/s3038842.htm.

Cullen, Simon. "Burke Unveils Final Murray–Darling Plan." Australian Broadcasting Corporation, November 22, 2012. http://www.abc.net.au/news/2012-11-22/burke-unveils-final-murray-darling-plan/4386298.

Davies, Richard. "Central China Floods 1931." Floodlist, April 10, 2013. http://floodlist.com/asia/central-china-floods-1931.

Davies, Richard. "120,000 Still Displaced by Flooding Rivers in Brazil, Argentina, Uruguay and Paraguay." Floodlist, January 8, 2016. http://floodlist.com/america/120000-displaced-floods-brazil-argentina-uruguay-paraguay.

Dodge, Jeff. "Researchers: Building Better Dams Starts with Ecological Insights." Fort Collins, CO: Colorado State University, September 2015. http://source.colostate.edu/researchers-building-better-dams-starts-with-ecological-insights/.

Environmental Working Group. "Dead in the Water." Washington, DC: April 2006. http://www.ewg.org/research/dead-water.

European Commission. "Flood Protection: Commission Proposes Concerted EU Action." Press release. Brussels, Belgium, July 12, 2004. http://europa .eu/rapid/press-release_IP-04-887_en.htm.

European Parliament. "Environment: European Parliament and Council Reach Agreement on the New Floods Directive." Press release. Brussels, Belgium, April 25, 2007. http://europa.eu/rapid/press-release_IP-07 -565_en.htm?locale=en.

Fears, Darryl. "U.S. Wildfires Just Set an Amazing and Troubling New Record." *Washington Post*, January 6, 2016. https://www.washingtonpost.com /news/energy-environment/wp/2016/01/06/2015-wildfire-season-just -set-an-amazing-and-troubling-new-record/?utm_term=.4070407012cf.

Fleck, John. "Post-Fire Runoff Led to 'Dead Zones.'" *Albuquerque Journal*, May 25, 2014. https://www.abqjournal.com/405929/postfire-runoff -led-to-dead-zones.html.

Fleck, John. "Rio Grande Water Restored." *Albuquerque Journal*, July 30, 2011. https://www.abqjournal.com/46645/rio-grande-water-restored-2.html.

Fountain, Henry. "Dry Winter and Warm Spring Set Stage for Wildfire in Canada." *New York Times*, May 5, 2016. https://www.nytimes.com/2016 /05/06/science/dry-winter-and-warm-spring-set-stage-for-canadian -inferno.html.

Friends of Georgica Pond Foundation. East Hampton, NY. http://friendsof georgicapond.org/.

Galbraith, Kate. "Texas Farmers Battle Ogallala Pumping Limits" *The Texas Tribune*, March 18, 2012. https://www.texastribune.org/2012/03/18 /texas-farmers-regulators-battle-over-ogallala/.

GE Water and Process Technologies. "Australian Golf Course Recycles Municipal Wastewater with Onsite ZeeWeed MBR." General Electric, June 2008. https://www.gewater.com/kcpguest/salesedge/documents /...Cust/.../CS1279EN.pdf.

Gila Conservation Coalition. Silver City, New Mexico. http://www.gilacon servation.org.

Gila Conservation Coalition, "ISC and NM CAP Entity Release Gila River Diversion Options." Silver City, NM: Spring 2016. http://www.gilacon servation.org/wp/?p=2275.

Gillis, Justin. "California Drought Is Made Worse by Global Warming, Scientists Say." *New York Times*, August 20, 2015. https://www.nytimes.com /2015/08/21/science/climate-change-intensifies-california-drought-sci entists-say.html.

Gillis, Justin. "Climate Chaos, Across the Map." *New York Times*, December 30, 2015. https://www.nytimes.com/2015/12/31/science/climate-chaos -across-the-map.html.

Gorder, Pam Frost. "Study Maps Hidden Water Pollution in U.S. Coastal Areas." Press release. Ohio State University, Columbus, August 4, 2016. https://news.osu.edu/news/2016/08/04/subwater/.

Greenspan, Jesse. "Remembering the Apollo 8 Christmas Eve Broadcast." *History*. http://www.history.com/news/remembering-the-apollo-8-christmas -eve-broadcast, posted November 23, 2015.

Guardian. "Last Winter's Floods 'Most Extreme on Record in UK,' Says Study." December 5, 2016. https://www.theguardian.com/environment/2016 /dec/05/last-winters-floods-most-extreme-on-record-uk-study.

Hellauer, Susan. "Sustainability Saturday: How the Rockland Desalination War Was Won." *Nyack News and Views*, May 7, 2016. http://www.nyack newsandviews.com/2016/05/susssat_shellauer_desal/.

Howard, Brian Clark. "Amid Drought, Explaining Colorado's Extreme Floods." September 13, 2013. http://news.nationalgeographic.com/news/2013 /09/130913-colorado-flood-boulder-climate-change-drought-fires/.

Howard, Brian Clark. "Saving the Colorado River Delta, One Habitat at a Time." December 15, 2014. http://news.nationalgeographic.com/news /special-features/2014/12/141216-colorado-river-delta-restoration -water-drought-environment/.

Hower, Mike. "Report: 91% of Top Firms View Climate Shocks as Business Risk." September 30, 2015. http://mail.sustainablebrands.com/news _and_views/ict_big_data/mike_hower/report_91_top_firms_view_cli mate_change_impacts_business_risk.

International Boundary and Water Commission. El Paso, Texas. https://www .ibwc.gov/.

International Commission for the Protection of the Danube River. "The Danube River Basin: Facts and Figures." Vienna, Austria. https://www.icpdr .org/main/sites/default/files/nodes/documents/icpdr_facts_figures.pdf.

International Commission on Large Dams. "World Register of Dams." http:// www.icold-cigb.org/GB/World_register/general_synthesis.asp.

International Desalination Association. "Desalination by the Numbers." http://idadesal.org/desalination-101/desalination-by-the-numbers/.

International Fertilizer Industry Association. "Nitrogen Fertilizer Nutrient Consumption." http://www.fertilizer.org/statistics.

International Water Management Institute. "First Global Estimate of Urban Agriculture Reveals Area Size of the EU That's Boosting Food Security in Cities." Press release. Colombo, Sri Lanka, November 13, 2014. http://www.iwmi.cgiar.org/2014/11/press-release-area-size-of-the-eu-thats-boosting-food-security-in-cities/.

Jacobs, Andrew. "Keep Your Mouth Closed: Aquatic Olympians Face a Toxic Stew." *New York Times*, July 26, 2016. https://www.nytimes.com/2016/07/27/world/americas/brazil-rio-water-olympics.html.

Jensen, Mari N. "Colorado River Delta Flows Help Birds, Plants, Ground-water." University of Arizona, College of Science, October 19, 2016. https://uanews.arizona.edu/story/colorado-river-delta-flows-help-birds-plants-groundwater.

Jha, Ravi S. "India's River-Linking Project Mired in Cost Squabbles and Politics." *Guardian*, February 5, 2013. https://www.theguardian.com/environment/2013/feb/05/india-river-link-plan-progress-slow.

Kendall, Clare. "A New Law of Nature." *Guardian*, September 23, 2008. https://www.theguardian.com/environment/2008/sep/24/equador.conservation.

Kerlin, Kat. "Flooding Farms in the Rain to Restore Groundwater." *UC–Davis News*, Davis, California, January 26, 2016. https://www.ucdavis.edu/news/flooding-farms-rain-restore-groundwater/.

Khew, Carolyn. "Fifth Singapore Desalination Plant in the Pipeline." *Straits Times*, July 13, 2016. http://www.straitstimes.com/singapore/environment/fifth-spore-desalination-plant-in-the-pipeline.

Killam, David G. "Sacramento District Project Wins Public Works Project of the Year." February 12, 2009. https://www.army.mil/article/16928/Sacramento_District_project_wins_Public_Works_Project_of_the_Year.

Kimmelman, Michael. "Going with the Flow." *New York Times*, February 13, 2013. http://www.nytimes.com/2013/02/17/arts/design/flood-control-in-the-netherlands-now-allows-sea-water-in.html.

Kimsey, Gary. "Rainstorm of 'Biblical Proportions' Teaches a Tough Lesson along the Poudre River." September 15, 2013. https://poudreriver.org

/2013/09/15/rainstorm-of-biblical-proportions-teaches-a-tough-lesson
-along-the-poudre-river/.

Kozacek, Codi. "Ethiopia Hunger Reaches Emergency Levels." Circle of Blue,
January 5, 2016. http://www.circleofblue.org/2016/world/ethiopia-hun
ger-reaches-emergency-levels/.

Leach, Anna. "Soak It Up: China's Ambitious Plan to Solve Urban Flooding with 'Sponge Cities'". *Guardian*, October 3, 2016. https://www.the
guardian.com/public-leaders-network/2016/oct/03/china-government
-solve-urban-planning-flooding-sponge-cities.

Liu, John D. "Lessons of the Loess Plateau." Environmental Education and
Media Project, June 2013. http://eempc.org/lessons-of-the-loess-plateau/.

Loomis, Brandon. "$20 Million Plan to Aid Arizona's Stressed-Out Verde River."
Arizona Republic, May 28, 2016. http://www.azcentral.com/story/news
/local/arizona-water/2016/05/28/arizona-verde-river-aid/83875824/.

Lustgarten, Abrahm. "Unplugging the Colorado River." *New York Times*,
May 20, 2016. https://www.nytimes.com/2016/05/22/opinion/unplug
ging-the-colorado-river.html.

Madera Irrigation District. "Madera Irrigation District Continues Water Deliveries and Massive Recharge Effort." Press release, Madera, California,
January 11, 2017. https://www.madera-id.org/index.php/145-press-re
lease-water-deliveries-and-recharge-effort.

Mankad, Rekha. "Does Grass-Fed Beef Have Any Heart-Health Benefits That
Other Types of Beef Don't?" Mayo Clinic, Cleveland, Ohio. http://www
.mayoclinic.org/diseases-conditions/heart-disease/expert-answers/grass
-fed-beef/faq-20058059.

Mapes, Linda V. "Elwha: Roaring Back to Life." *Seattle Times*, February 13,
2016. http://projects.seattletimes.com/2016/elwha/.

McWilliams, James E. "All Sizzle and No Steak: Why Allan Savory's TED
Talk about How Cattle Can Reverse Global Warming Is Dead Wrong."
Slate, April 22, 2013. http://www.slate.com/articles/life/food/2013/04
/allan_savory_s_ted_talk_is_wrong_and_the_benefits_of_holistic_graz
ing_have.html.

Melbourne Water. https://www.melbournewater.com.au.

Miller, Matt. "Building Wetlands for Clean Drinking Water." The Nature Conservancy, January 30, 2013. http://blog.nature.org/science/2013/01/30
/feature-building-wetlands-for-clean-water/.

Milman, Oliver. "Florida Declares State of Local Emergency over Influx of 'God-awful' Toxic Algae." *Guardian*, June 30, 2016. https://www.the guardian.com/us-news/2016/jun/30/florida-emergency-toxic-algae -treasure-coast.

Mississippi River/Gulf of Mexico Hypoxia Task Force. "Hypoxia Task Force Success Stories." Washington, DC: US Environmental Protection Agency. https://www.epa.gov/ms-htf/hypoxia-task-force-success-stories.

Mississippi River Trust. "Restoring Mississippi River Bottomland Hardwood Forests." http://www.mississippirivertrust.org/.

Moebius-Clune, Bianca. "Introducing the New Soil Health Division." US Department of Agriculture, Natural Resources Conservation Service, webinar, January 12, 2016. http://www.forestrywebinars.net/webinars /the-new-division-of-soil-health-approach-and-benefits/.

Monterey Regional Water Pollution Control Agency. "Turning Wastewater into Safe Water." http://www.mrwpca.org/about_facilities_water_recycling.php.

National Geographic Society. "Water Footprint Calculator, Methodology and Tips." http://environment.nationalgeographic.com/environment/fresh water/water-calculator-methodology/.

National Public Radio. "As Brazil's Largest City Struggles with Drought, Residents Are Leaving." November 22, 2015. http://www.npr.org/sections /parallels/2015/11/22/455751848/as-brazils-largest-city-struggles-with -drought-residents-are-leaving.

National Public Radio. "Saudi Hay Farm in Arizona Tests State's Supply of Groundwater." November 2, 2015. http://www.npr.org/sections/the salt/2015/11/02/453885642/saudi-hay-farm-in-arizona-tests-states-sup ply-of-groundwater.

New York City Department of Environmental Protection, "Land Acquisition." http://www.nyc.gov/html/dep/html/watershed_protection/land_acquisi tion.shtml.

New York State Department of Environmental Conservation. "Harmful Algal Blooms and Marine Biotoxins." Albany, NY. http://www.dec.ny.gov/out door/64824.html.

Opperman, Jeff. "Fish Run through It: The Importance of Maintaining and Reconnecting Free-Flowing Rivers." *Water Currents*, May 18, 2016. http://voices.nationalgeographic.com/2016/05/18/fish-run-through-it -the-importance-of-maintaining-and-reconnecting-free-flowing-rivers/.

Our Town. "Leaks Are Us: United Water's Fuzzy Data." Op-ed., July 1, 2015.

Parker, Laura. "Slimy Green Beaches May Be Florida's New Normal." National Geographic, July 27, 2016. http://news.nationalgeographic.com/2016/07/toxic-algae-florida-beaches-climate-swamp-environment/.

Postel, Sandra. "Arizona Irrigators Share Water with Desert River." *Water Currents*, September 3, 2013. http://voices.nationalgeographic.com/2013/09/03/arizona-irrigators-share-water-with-desert-river/.

Postel, Sandra. "A Dam, Dying Fish, and a Montana Farmer's Lifelong Quest to Right a Wrong." *Water Currents*, September 5, 2013. http://voices.nationalgeographic.com/2013/09/05/a-dam-dying-fish-and-a-montana-farmers-lifelong-quest-to-right-a-wrong/.

Postel, Sandra. "Fire and Rain: The One-Two Punch of Flooding after Blazes." *Water Currents*, August 31, 2011. http://voices.nationalgeographic.com/2011/08/31/fire-and-rain/.

Postel, Sandra. "How Smarter Irrigation Might Save Rare Mussels and Ease a Water War." *Water Currents*, August 19, 2016. http://voices.nationalgeographic.com/2016/08/19/how-smarter-irrigation-might-save-rare-mussels-and-ease-a-water-war/.

Postel, Sandra. "How the Yampa River, and Its Dependents, Survived the Drought of 2012." *Water Currents*, October 18, 2012. http://voices.nationalgeographic.com/2012/10/18/how-the-yampa-river-and-its-dependents-survived-the-drought-of-2012/.

Postel, Sandra. "Lessons from São Paulo's Water Shortage." *Water Currents*, March 13, 2015. http://voices.nationalgeographic.com/2015/03/13/lessons-from-sao-paulos-water-shortage/.

Postel, Sandra. "A River in New Zealand Gets a Legal Voice." *Water Currents*, September 4, 2012. http://voices.nationalgeographic.com/2012/09/04/a-river-in-new-zealand-gets-a-legal-voice/.

Postel, Sandra. "Solar Electricity Buybacks May Reduce Groundwater Depletion in India." *Water Currents*, June 19, 2015. http://voices.nationalgeographic.com/2015/06/19/solar-electricity-buybacks-may-reduce-groundwater-depletion-in-india/.

Postel, Sandra. "That Sinking Feeling about Groundwater in Texas," *Water Currents*, July 19, 2012. http://voices.nationalgeographic.com/2012/07/19/that-sinking-feeling-about-groundwater-in-texas/.

Postel, Sandra. "Two Arizona Vineyards Give Back to a River through a Voluntary Water Exchange." *Water Currents*, July 25, 2016. http://voices
.nationalgeographic.com/2016/07/25/two-arizona-vineyards-give-back
-to-a-river-through-a-voluntary-water-exchange/.

PUB (Singapore's National Water Agency). https://www.pub.gov.sg/.

Richtel, Matt, and Fernanda Santos. "Wildfires, Once Confined to a Season, Burn Earlier and Longer." *New York Times*, April 12, 2016. https://www
.nytimes.com/2016/04/13/science/wildfires-season-global-warming
.html.

Robbins, Jim. "Deforestation and Drought." *New York Times*, October 9, 2015. https://www.nytimes.com/2015/10/11/opinion/sunday/deforesta
tion-and-drought.html.

Robbins, Jim. "In Napa Valley, Future Landscapes Are Viewed in the Past." *New York Times*, January 25, 2016. https://www.nytimes.com/2016/01/26
/science/in-napa-valley-future-landscapes-are-viewed-in-the-past.html.

Robertson, Campbell, and Alan Blinder. "As Louisiana Floodwaters Recede, the Scope of Disaster Comes into View." *New York Times*, August 16, 2016. https://www.nytimes.com/2016/08/17/us/louisiana-flooding.html.

Rocha, Jan. "Drought Bites as Amazon's 'Flying Rivers' Dry Up." Climate News Network, September 14, 2014. http://climatenewsnetwork.net
/drought-bites-as-amazons-flying-rivers-dry-up/.

Rockland County, Office of the County Executive. Press release. New City, New York, June 27, 2016. http://rocklandgov.com/departments/county
-executive/press-releases/2016-press-releases/rockland-files-lawsuit
-against-water-company-state-agencies/.

Rockland County, Office of the Legislature. "Independent Report: Millions of Gallons of Water Available to Rockland County—Rockland Task Force Files Major Water Study with State." Press release. New City, New York, July 28, 2015.

Rockland Water Coalition, http://www.sustainablerockland.org/.

Rockland Water Coalition. "As PSC Decision Looms on Rockland Water Rate Hikes and a Flawed Conservation Plan, Residents Question Whether a Private Corporation Should Manage Public Water." Media Alert. December 23, 2016.

Rodriguez, Robert. "Farmers and Water Districts Hope Storm Runoff Can

Help Replenish Underground Supplies." *Fresno Bee*, January 10, 2017. http://www.fresnobee.com/news/local/article125762529.html.

Rogoway, Mike. "Apple's Newest Innovation: Wastewater Treatment to Cool Prineville Data Centers." *The Oregonian*, June 12, 2016. http://www.ore gonlive.com/silicon-forest/index.ssf/2016/06/apples_newest_innova tion_waste.html.

Schlossberg, Tatiana. "Climate Change Will Bring Wetter Storms in U.S., Study Says." *New York Times*, December 6, 2016. https://www.nytimes .com/2016/12/06/science/global-warming-extreme-storms.html.

Schneider, Keith. "In Water-Scarce Regions Desalination Plants Are Risky Investments." Circle of Blue, November 22, 2016. http://www.circleofblue. org/2016/asia/water-scarce-regions-desalination-plants-risky-investments/.

Schneider, Keith. "Popularity of Big Hydropower Projects Diminishes around the World." Circle of Blue, July 14, 2016. http://www.circleofblue.org /2016/water-energy/hydropower/popularity-big-hydropower-projects -diminishes-around-world/.

Schwartz, John. "Water Flowing from Toilet to Tap May Be Hard to Swallow." *New York Times*, May 8, 2015. https://www.nytimes.com/2015/05/12 /science/recycled-drinking-water-getting-past-the-yuck-factor.html.

Scripps Institution of Oceanography. "Lake Mead Could Be Dry by 2021." University of California–San Diego, February 12, 2008. https://scripps .ucsd.edu/news/2487.

Semple, Kirk. "Long Island Sees a Crisis as It Floats to the Surface." *New York Times*, June 5, 2015. https://www.nytimes.com/2015/06/06/nyregion /long-island-sees-a-crisis-as-it-floats-to-the-surface.html.

Shabecoff, Philip. "Global Warming Has Begun, Expert Tells Senate." *New York Times*, June 24, 1988. http://www.nytimes.com/1988/06/24/us/global -warming-has-begun-expert-tells-senate.html?pagewanted=all.

Shepard, Wade. "Can 'Sponge Cities' Solve China's Urban Flooding Problem?" Citiscope, July 28, 2016. http://citiscope.org/story/2016/can-sponge -cities-solve-chinas-urban-flooding-problem.

Shuttleworth, Kate. "Agreement Entitles Whanganui River to Legal Identity." *New Zealand Herald*, August 30, 2012. http://www.nzherald.co.nz/nz /news/article.cfm?c_id=1&objectid=10830586.

Singler, Amy. "If You Remove It They Will Come—Restoring Amethyst Brook, MA." The River Blog, American Rivers, July 19, 2013.

Sonoran Institute. "From Sewage to Sanctuary: Las Arenitas Treatment Wetland." Tucson, Arizona. https://sonoraninstitute.org/card/from-sewage-to-sanctuary-las-arenitas-treatment-wetland/.

Sonoran Institute. "Restoring Mexico's Colorado River Delta, One Tree at a Time." Media Alert. Tucson, Arizona, October 17, 2016. https://sonoraninstitute.org/2016/restoring-mexicos-colorado-river-delta-one-tree-at-a-time/.

Southern Nevada Water Authority, Las Vegas, NV. https://www.snwa.com/.

Steinbock, Dan. "High Time to Reduce Costs of Floods." *China Daily*, July 7, 2016. http://usa.chinadaily.com.cn/opinion/2016-07/27/content_26235523.htm.

Stockholm International Water Institute. "Rajendra Singh—the Water Man of India Wins 2015 Stockholm Water Prize." Press release and interview, Stockholm, Sweden, March 20, 2015. http://www.siwi.org/prizes/stockholmwaterprize/laureates/2015-2/.

Strom, Stephanie. "Cover Crops, a Farming Revolution with Deep Roots in the Past." *New York Times*, February 6, 2016. https://www.nytimes.com/2016/02/07/business/cover-crops-a-farming-revolution-with-deep-roots-in-the-past.html.

Tang, Alisa. "From Open Sewage to High-Tech Hydrohub, Singapore Leads Water Revolution." Reuters, August 2, 2015. http://www.reuters.com/article/us-singapore-water-idUSKCN0Q804T20150803.

Tarun Bharat Sangh. http://tarunbharatsangh.in/.

The Nature Conservancy. "The Conservancy Reconnected Miles of Floodplain Forest Back to Louisiana's Ouachita River." https://www.nature.org/ourinitiatives/habitats/riverslakes/explore/largest-floodplain-restoration-in-mississippi-river-basin.xml?redirect=https-301.

The Nature Conservancy. "Modernizing Water Management: Building a National Sustainable Rivers Program." https://www.nature.org/ourinitiatives/habitats/riverslakes/sustainable-rivers-project.xml.

United Nations Children's Fund (UNICEF). "Malnutrition Mounts as El Niño Takes Hold." Johannesburg/Nairobi, Kenya, February 17, 2016.

United Nations Food and Agriculture Organization. Aquastat Database. http://www.fao.org/nr/water/aquastat/main/index.stm.

University of California–Irvine. "Parched West Is Using Up Underground Water, UCI, NASA Find." Press release, Irvine, California, July 24, 2014.

https://news.uci.edu/press-releases/parched-west-is-using-up-under
ground-water-uci-nasa-find/.

University of Illinois, College of Agricultural, Consumer and Environmental Sciences. "Illinois River Water Quality Improvement Linked to More
Efficient Corn Production." Urbana, Illinois, May 11, 2016. http://www
.sciencenewsline.com/news/2016051115230061.html.

University of Illinois, College of Agricultural, Consumer and Environmental
Sciences. "Wetlands Continue to Reduce Nitrates." Urbana, Illinois, May
12, 2015. http://www.sciencenewsline.com/news/2015051321150079
.html.

University of Maine, the Penobscot River Restoration Trust, and NOAA.
"After More than a Century, Endangered Shortnose Sturgeon Find Historic Habitat Post Dam Removal," November 16, 2015. https://umaine
.edu/news/blog/2015/11/16/after-more-than-a-century-endangered
-shortnose-sturgeon-find-historic-habitat-post-dam-removal/.

US Bureau of Reclamation. "Lower Colorado River Operations." Washington, DC: US Department of the Interior. https://www.usbr.gov/lc/region
/g4000/hourly/mead-elv.html.

US Department of Agriculture, Natural Resources Conservation Service. "Soil
Quality for Environmental Health." http://soilquality.org/indicators
/total_organic_carbon.html.

US Executive Office of the President. "Taking Action to Protect Communities and Reduce the Cost of Future Flood Disasters." Fact Sheet. Washington, DC, January 30, 2015. https://www.fema.gov/media-library
-data/1422641422398-ac287820c1d571f07c0d585a9f593af3/15-01
-30-flood-standards-fact-sheet.pdf.

US Geological Survey. "Irrigation Causing Declines in the High Plains Aquifer." Press release. Reston, Virginia, February 1, 2012. https://archive.usgs
.gov/archive/sites/www.usgs.gov/newsroom/article.asp-ID=3093.html.

US National Aeronautics and Space Administration (NASA). "Apollo 8:
Christmas at the Moon." https://www.nasa.gov/topics/history/features
/apollo_8.html, posted December 18, 2014.

US National Aeronautics and Space Administration (NASA). Earth Observa-
tory. https://earthobservatory.nasa.gov/.

US National Oceanic and Atmospheric Administration (NOAA). "Average 'Dead Zone' for Gulf of Mexico Predicted." Silver Spring, MD: US Department of Commerce, June 9, 2016. http://www.noaa.gov/media -release/average-dead-zone-for-gulf-of-mexico-predicted.

US National Oceanic and Atmospheric Administration (NOAA). "NOAA and Partners Cancel Gulf Dead Zone Summer Cruise." Silver Spring, MD: US Department of Commerce, July 29, 2016. http://www.noaa.gov /media-release/noaa-and-partners-cancel-gulf-dead-zone-summer-cruise.

US National Oceanic and Atmospheric Administration (NOAA). "NOAA-Supported Scientists Find Large Gulf Dead Zone, but Smaller than Predicted." Silver Spring, MD: US Department of Commerce, July 29, 2013. http://www.noaanews.noaa.gov/stories2013/2013029_deadzone .html.

US National Oceanic and Atmospheric Administration (NOAA). "2016 Marks Three Consecutive Years of Record Warmth for the Globe." Silver Spring, MD: US Department of Commerce, January 18, 2017. http:// www.noaa.gov/stories/2016-marks-three-consecutive-years-of-record -warmth-for-globe.

US National Oceanic and Atmospheric Administration (NOAA). National Weather Service, Hydrometeorological Design Studies Center. "Exceedance Probability Analysis for Selected Storm Events." Silver Spring, MD: US Department of Commerce. http://www.nws.noaa.gov/oh/hdsc/aep _storm_analysis/.

Vastag, Brian, and Frances Stead Sellers. "Floods along the Mississippi River Lead to Renewed Calls for a Change in Strategy." *Washington Post*, May 8, 2011. https://www.washingtonpost.com/national/floods-along-the-mis sissippi-river-lead-to-renewed-calls-for-a-change-in-strategy/2011/05/05 /AFjngLUG_story.html?utm_term=.d8a3a06b3b14.

Vidal, John. "Bolivia Enshrines Natural World's Rights with Equal Status for Mother Earth." *Guardian*, April 10, 2011. https://www.theguardian .com/environment/2011/apr/10/bolivia-enshrines-natural-worlds-rights.

Vidot, Anna, and Brett Worthington. "Murray–Darling Basin Authority Recommends Reducing Water Buybacks in Northern Communities." Australian Broadcasting Corporation, November 22, 2016. http://www .abc.net.au/news/2016-11-22/murray-darling-basin-northern-commu nities/8042496.

Water Footprint Network. "Product Gallery." The Hague, the Netherlands. http://waterfootprint.org/en/resources/interactive-tools/product-gallery/.

WaterWorld. "Kansas City Water's Green Infrastructure Projects Win Sustainability Award." Washington, DC: October 11, 2016. http://www.water world.com/articles/2016/10/kc-water-s-green-infrastructure-projects -win-sustainability-award.html.

Wendycitychicago.com. "Chicago Green: Roofs." March 20, 2015. http:// wendycitychicago.com/?s=green+roof.

Wines, Michael. "Behind Toledo's Water Crisis, a Long Troubled Lake Erie." *New York Times*, August 4, 2014. https://www.nytimes.com/2014/08/05 /us/lifting-ban-toledo-says-its-water-is-safe-to-drink-again.html?_r=0.

Wisnieski, Adam. "City's Watershed Protection Plan Seeks Difficult Balance Upstate." *City Limits*, June 15, 2015. http://citylimits.org/2015/06/15 /citys-watershed-protection-plan-seeks-difficult-balance-upstate/.

World Tibet Network News. "Death Toll in China Sandstorm Rises to 47." May 11, 1993. http://www.tibet.ca/en/library/wtn/archive/old?y=1993 &m=5&p=11_4.

Yolo Basin Foundation. "About the Yolo Bypass Wildlife Area." http://yoloba sin.org/yolobypasswildlifearea/.

Index

Figures/photos/illustrations are indicated by an "f."